Circuit Analysis

The direction in which education starts a man will determine his future life
Plato: *The Republic, Book V* (427-347 BC)

Talking of education, people have now a-days (said he) got a strange opinion that every thing should be taught by lectures. Now, I cannot see that lectures can do so much good as reading the books from which the lectures are taken. I know nothing that can be best taught by lectures, expect where experiments are to be shewn. You may teach chymestry by lectures. — You might teach making of shoes by lectures!
James Boswell: *Life of Samuel Johnson 1766* (1709-1784)

JOHN E. WHITEHOUSE, B.Sc., Ph.D.

John Whitehouse is Lecturer in the Department of Engineering at the University of Reading. Educated at Ilford County High School in Essex, he then graduated at the University of Reading with First Class Honours in physics (1951), and a Ph.D. (1961) for his thesis on scattering and emission of light from quartz and diamond. He worked in the USA at Purdue University as Research Associate for nearly five years, studying the effects of high energy radiation on semiconductors at very low temperatures, which work was published in *Physical Review*. He returned to England in 1965 to a post in his old department in Reading University and continued work on semiconductors, and organised and subsequently edited the proceedings of the International Conference on Radiation Damage in Semiconductors (1973).

As Lecturer in Physics John Whitehouse teaches a variety of topics, often in association with departments of Chemistry and Mathematics. He taught solid state physics and also a course on electromagnetism, which was attended by engineering students and applicable in electronics for physicists. In 1984 he transferred to the Department of Engineering where he added to his teaching courses one on circuit analysis and, more recently, another on digital signal processing. In his present department he has served as Admissions Tutor and is now Examinations Officer.

His current research on electromagnetic compatibility (EMC) stems from his many years of teaching electromagnetism. He was recently named as joint inventor in the patent on a cell of novel design for use in EMC testing (Web page www.elec.reading.ac.uk/eme.html).

His pastimes are railway signalling and restoring Indian carpets. He just likes **doing** things (there are few devices that he cannot fix or try to build, e.g. bricklayer, carpenter, plumber, electrician). He has travelled much in the USA and continental Europe.

Circuit Analysis

John E. Whitehouse, B.Sc., Ph.D., C.Eng., M.I.E.E.
Department of Engineering
University of Reading

Horwood Publishing
Chichester

Published in 1997 by
HORWOOD PUBLISHING LIMITED **(formerly ALBION PUBLISHING)**
International Publishers
Coll House, Westergate, Chichester, West Sussex, PO20 6QL
England

British Library Cataloguing in Publication Data
A catalogue record of this book is available from the British Library

ISBN 1-898563-40-3

Printed and bound in Great Britain by MPG Books Ltd, Bodmin, Cornwall

Contents

Author's Preface ix

1 Fundamentals   **1**
    1.1   The Electronic Engineer 1
    1.2   The Electronic System 1
    1.3   The Linear System 2
    1.4   Passive and Active Circuits 3
    1.5   Ideal Sources and Practical Sources 5
    1.6   Stability and Causality 11
    Summary 13
    Problems 13

2 Network Equations   **15**
    2.1   Kirchhoff's Laws 15
    2.2   The Node and Loop Equations 18
    2.3   Tellegen's Theorem 23
    2.4   The State Equations 26
    Summary 27
    Problems 28

3 Network Theorems   **30**
    3.1   Superposition in Network Analysis 30
    3.2   Thevenin's and Norton's Theorems 32
    3.3   Power Transfer between Systems 39
    3.4   Network Duals 40
    Summary 42
    Problems 43

4 Networks with Inductors and Capacitors   **45**
    4.1   General n-terminal and Two-port Networks 45
    4.2   The Free Response of Systems 46
    4.3   The Role of the Sine Function 51
    4.4   Response to a Simple Harmonic Input 54
    4.5   Mutual Inductance: Networks with Transformers .. 57
    Summary 60
    Problems 61

5 Network Analysis using Phasors   **63**
   5.1 Complex Exponential Representation of Simple Harmonic Motion   63
   5.2 A General Solution for the Free Response   68
   5.3 The Complex Frequency   69
   5.4 Response to a Complex Exponential Input   70
   5.5 The nth Order System   73
   5.6 The Complex Impedance   74
   5.7 Networks with Transformers   77
   Summary   77
   Problems 79

6 The Laplace Transform in Network Analysis   **80**
   6.1 Operational Calculus   80
   6.2 The Laplace Transform   81
   6.3 Application to Network Equations   84
   6.4 The General Interpretation of the System Function   85
   6.5 The Laplacian Impedance   86
   6.6 Finding the Time Domain Response   89
   6.7 The Impulse Response of Networks   94
   6.8 Non-zero Initial Stored Energy   97
   6.9 The Bilateral Laplace Transform   98
   Summary   99
   Problems 100

7 The Fourier Series and Fourier Transform   **102**
   7.1 Time and Frequency Domains   102
   7.2 Fourier's Theorem   103
   7.3 The Fourier Transform   109
   7.4 Impulses in Time and Frequency: Duality and Convolution 111
   7.5 Time-shifted Impulses and Impulse Sequences   114
   7.6 Rectangular Pulses in Time and Frequency   117
   Summary   120
   Problems 122

8 The Frequency Response of Networks   **124**
   8.1 The Response to Steady a.c.   124
   8.2 The Steady a.c. Impedance   126
   8.3 Driving-point Impedance and Admittance   129
   8.4 Reflected Impedance in Networks with Transformers   135
   8.5 Logarithmic Scales, Decibels and Bode Plots   137
   8.6 First-order Filters 139
   8.7 Second-order Band-pass Filters   143
   8.8 The Quality Factor, Q   146
   8.9 Second-order Notch, Low-pass and High-pass Filters   148
   Summary   151
   Problems 153

9 Power Dissipation and Energy Storage in Networks   **155**
   9.1   Power Dissipation in Resistive Networks 155
   9.2   Power Dissipation in Networks with Reactive Components 157
   9.3   The Complex Power   161
   9.4   Stored Energy in an LCR Circuit 162
   Summary   164
   Problems   165

A Coefficients in the Fourier Series   **167**

B General Solution of the First-order Linear Differential Equation   **169**

C Laplace Operational Transforms   **171**

D Recommended Reading   **173**

E Answers and Guidance for Problems   **174**

Index   **180**

Author's Preface

The present text is a successor to my earlier *Principles of Network Analysis*, Ellis Horwood, 1991. What may appear to be an inconsequential change is the title of the text. Here I have taken the advice of the publishers and moved from the technically more correct 'Network Analysis' to the more colloquial 'Circuit Analysis'. So closely is the word *network* associated with computers that the former title caused confusion in regard to cataloguing. Not for the first time in response to 'what's in a name?' one hears 'quite a lot!'

The impetus to produce the book was, as must so often be the case, the need to support a course of lectures given by the author. The course in this particular case was an introductory series of lectures on 'Circuit Theory' given in the first two terms of a three-year BEng course in Electronic Engineering. The Laplace transform and Fourier analysis are now studied together in a separate course after that on elementary circuit theory. The treatment of the use of transform methods in the book has been extended to cater for such a course. The book is therefore now appropriate for use in courses which in most institutions will span the first two years of the degree.

Every text reflects what the author believes to be his, or her, unique perspective on the subject. That is true here. I can only say that I have been dissatisfied with the presentation elsewhere of how one should take the next of each of the various hurdles, or indeed, why it should be taken at all! I have tried to 'come clean' with the students, explaining how, on occasion, they are having to hear about a particular topic but may, in reasonable conscience, then put it to the back of their minds. The role of differential equations in network analysis is a case in point; it must be understood that they govern the behaviour of all physical systems, of which networks are just an example, and we hope students will retain a competence in solving them, but we are more concerned that the method of impedances, which bypasses this step, will make an indelible impression.

The intention has been to follow the thread of network analysis in as economical a way as possible. Worked examples in the text have been kept to a necessary minimum in the expectation that teachers will generally want to devise their own tutorial support. There is otherwise a risk that one may not 'see the wood for the trees'. This is therefore very much a 'theme text' rather than a 'workbook' and its attraction may be that it is compact. As to how compact, I can accept the possibility that it is sufficiently succinct as to need reading and re-reading but I hope it is not succinct to the point of being cryptic. It could be hoped that a keen student might read the whole book in a few sittings without pausing to work through problems but then on a second pass would see the need to develop expertise in particular areas by means of problem-working.

A change in this edition, in itself quite minor, is the adoption of the BSI standard symbol for a voltage source. The symbol shows, using a + sign, the terminal towards which the reference direction for current is directed. Abandoning the use of the voltage arrow, which indicates the direction of voltage increase, has provided the opportunity to emphasize instead voltage drops across components and sources. The voltage drop is in the same sense as the reference direction for current and so a single arrow may now be used to give the sense of both the current and the voltage change. It follows that when we draw on the results of graph theory to help with a statement of Kirchhoff's laws in matrix form a single directed graph is sufficient to represent a circuit. Where it is felt necessary to draw attention to a voltage *difference* either across a component or between two terminals, a pair of arrows is quite intentionally used. It may require a solution of the circuit equations to decide the sense of the particular voltage change. The overall effect of the adoption of new conventions has therefore, far from being minor, been quite major. It is hoped that lecturers already employing the text in their taught courses will see the changes as a useful simplification and be able to adapt their teaching material without huge effort. The 'figure-of-eight' symbol for the current source has been left unchanged. It is very close to the BSI symbol except that, by personal preference, the current is shown by an arrow head *on* the conductor rather than in the space around the source.

John E. Whitehouse

Reading, August 1997

1

Fundamentals

1.1 The Electronic Engineer

The education of all engineers seeking professional qualification prepares them to perform original and creative work. Professional standing and what it entails should be the goal of any student embarking on a course based upon this text. The professional electronic engineer, in particular, designs the circuits and systems which, *ipso facto*, are not yet invented. In contrast, students at this point in their careers are often already enthusiastic constructors of circuits culled from the popular electronic press and so originated by someone else. Proficiency in circuit construction is a splendid, but not essential, background to an electronic engineering course. To become originators of circuits the intending professional must acquire an entirely new set of skills. The purpose of this text is to introduce the most basic of those skills.

Before new circuits may be designed, a thorough understanding is required of the fundamental principles of operation of all circuits. That understanding is gained from an analysis of the operation of circuits. Our motto is therefore: Before we may synthesize we must first analyze. Not surprisingly, we shall confine our attention on each occasion to the simplest circuit which illustrates the principle under consideration. Finally on this point, an untaught skill must also be developed, namely a degree of scepticism, one part of common sense. It is required, for instance, as a check when approaching a particular design that all of the rules are in fact in one's possession. An effort will be made here to point to as many areas as possible which contribute to a successful design even though, owing to limitations of space, some cannot be treated in any great depth.

1.2 The Electronic System

Any combination of interconnected electrical components constitutes a circuit (or network). It is one example of a physical system. When such a system is subject to an input (excitation) it produces an output (response) (Figure 1.1). The relationship of the output to the input is given by the system function, and from our point of view this is the most important property of the system. A particular system function could be achieved in any number of ways, the choice of circuit will be determined

1

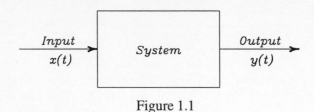

Figure 1.1

by aesthetics (neatness) and economics (component cost). This exemplifies the systems approach to electronics. For ease of analysis a system may be broken down into sub-systems, each of which will be characterized by a system function. In almost every case our analysis will be at the sub-system level.

1.3 The Linear System

If for a particular input $x_1(t)$ we have the output $y_1(t) = f\{x_1(t)\}$ and for the input $x_2(t)$ we have the output $y_2(t) = f\{x_2(t)\}$ then for the input $\{x_1(t) + x_2(t)\}$ we could in general expect an output $f\{x_1(t) + x_2(t)\}$. The output for the sum of inputs would bear no simple relationship to the output for the individual inputs (Figure 1.2). However, we can readily suggest a very simple way in which the output for a sum of inputs may be related to the outputs for the individual inputs and which it might be reasonable to suppose obtains for many circuits. The output that is the sum of the outputs for the individual inputs would be the simplest case. A circuit which behaves in this way is said to be a *linear* circuit and obeys the *principle of superposition*. All other circuits are non-linear. Nature makes a great concession to us in allowing that very many physical systems behave linearly. Obviously no real device can continue to behave linearly for an ever-increasing input (Figure 1.3). Here then is an implicit rule, linear within limits. Linearity does not simply mean relationships of the sort $y = ax$, as we may see in the following example.

Example 1.1 The input $x_1(t)$ to a differentiator will produce the output:

$$y_1(t) = \frac{dx_1(t)}{dt}$$

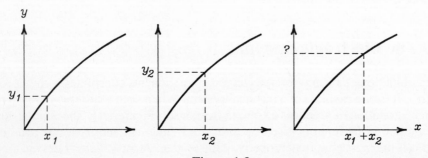

Figure 1.2

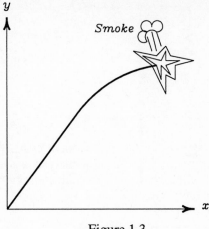

Figure 1.3

while the input $x_2(t)$ will produce:

$$y_2(t) = \frac{dx_2(t)}{dt}$$

The inputs applied together will lead to an output:

$$\frac{d\{x_1(t) + x_2(t)\}}{dt} = \frac{dx_1(t)}{dt} + \frac{dx_2(t)}{dt}$$
$$= y_1(t) + y_2(t)$$

that is, the sum of the outputs for the inputs applied separately. The differentiator is therefore a linear system.

More generally, linearity is not expressed simply in terms of the sum of inputs and the sum of outputs but as a *linear combination* in each case. We say that for the input $K_1x_1(t) + K_2x_2(t)$ we expect the output $K_1y_1(t) + K_2y_2(t)$ where K_1 and K_2 are constants.

1.4 Passive and Active Circuits

We now distinguish two important classes of circuit:

(i) A passive circuit is one in which there is a net dissipation of energy. It contains the passive circuit elements resistors, capacitors and inductors. The energy loss occurs by joule heating in the resistors.

(ii) An active circuit is one which can supply energy to other circuits. In addition to resistors, capacitors and inductors, it will contain such components as transistors. The active element in each case includes its power supply which in its simplest form may be just a battery.

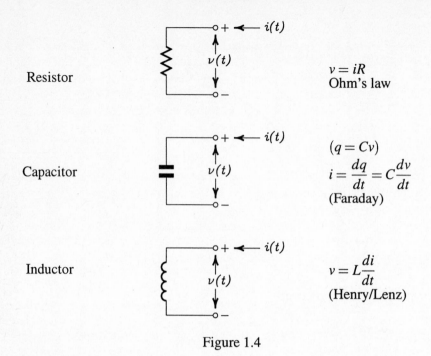

Figure 1.4

Ideal resistors, capacitors and inductors have simple equations connecting the voltage applied to them and the current through them. Treating the current here in the classical manner, as a flow of positive charges, the current–voltage relationships for the three cases are set out in Figure 1.4 where we have also adopted the *passive sign convention*. In that convention the reference direction for current is *into* the terminal at the highest positive potential and a current in that direction is considered to be positive. The convention accords with the definition of a passive circuit element given above which requires that a positive charge delivered to the more positive terminal will fall in energy as it progresses to the terminal at the lower potential. For a resistor, the energy lost will be dissipated as joule heating. For a capacitor or an inductor the current will initially increase the fields which temporarily store energy but this energy is given up when the source is turned off or reversed in polarity. In addition to defining the reference direction for current, it will also be convenient for what follows in later chapters to attach a direction to the *sense* of the voltage change across a component. We may attach significance to either the direction of the voltage drop or the directon of the voltage rise. Both choices have been adopted in other treatments. In all that follows we will work with the direction of the *voltage drop*, which is in the same sense as the reference direction for current in the component. A single arrow will therefore suffice to characterize the sense of both the voltage and the current in a passive component and this provides a useful simplification in our later work.

For resistors the current–voltage relationship shown in Figure 1.4 is Ohm's law and the ohm (Ω) is the unit for the constant of proportionality. For capacitors and inductors the units for the constants are the farad and the henry respectively, derived from the names of Faraday and Henry although the name of Lenz is usually attached to the expression for the inductor. The three expressions are our first simple differential equations. It will be seen that time dependence is of the essence. A real resistor will have some inductance associated with it, as will even a simple length of wire, and possibly also some stray capacitance. A real capacitor has some series resistance and inductance in its leads and further shunt resistance representing the leakage of its dielectric. A real inductance similarly has series resistance and stray capacitance. It may be necessary, particularly at high frequencies, to show these added contributions explicitly.

Implicit in our description of the three simple components is the idea of the 'lumped parameter', itself central to the application of circuit theory. The frequencies at which the circuits we are studying will be employed are such that spatial variations of voltage and current over a component will be negligible. We can then assign a unique value to each electrical component valid at all frequencies of interest. At microwave frequencies this is not so and field theory rather than circuit theory is required.

1.5   Ideal Sources and Practical Sources

Active circuits, as we have seen, necessarily include a battery or power supply. The latter converts a 50 or 60 Hz electrical supply to d.c. but does so in a very nonlinear way. Apart from their use in powering active circuits, power supplies are outside our area of interest. They may be thought of very much as batteries except that certain protections against excess current and voltage are often built-in. Apart from the power for active circuits, inputs are also made to these and passive circuits from a host of different transducers, for instance radio antennae, microphones or photocells, and also from other active circuits, for instance a pre-amplifier feeding a power amplifier. While there is a unique relationship between the current and the applied voltage for a passive component, the relationship between the output voltage and current for a battery, a power supply, a transducer or an active circuit is determined by the characteristics of the network to which they are connected. We identify all such components with this property as *sources*.

In order to develop models for actual sources we proceed in the same way whether dealing with power supplies or signal sources but just bear in mind that as signal sources are functions of time we have to confine our attention to a particular instant. We first introduce the concept of *ideal sources*. An *ideal independent source* can maintain a particular voltage or current whatever the load upon it and whatever the circumstances in the rest of the circuit. An *ideal dependent (or controlled) source* has the same properties as regards the load but, in addition, depends on an internal state (current or voltage) of the system. Controlled sources allow us to model the active elements of a network, for instance a transistor where the collector current is controlled by the base-emitter voltage (Figure 1.5). Since an active (control) element includes its power supply it follows that an independent source is implicit in the notion of a controlled source. The controlled source is a means of representing at one pair of terminals what is in essence a three-terminal element. Ideal sources are by definition an abstraction and cannot be realized as isolated elements, although some practical sources are close to being ideal, for instance, an ordinary car battery. The symbols we use for ideal independent voltage

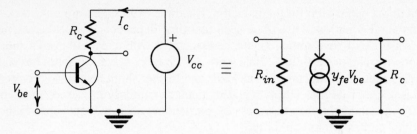

Figure 1.5

and current sources are shown in Figure 1.6 together with their current–voltage
$(i - v)$ characteristics, where the reference direction for current in each case is the
same as that used for passive components, that is, current *into* the most positive
terminal is taken as positive. The ideal independent voltage sources will naturally
be able to deliver a current to an external passive circuit so that the physical current
of a source will be in the opposite direction to the reference and therefore always
assigned a *negative* sign. The same is true for the ideal independent current source.
Following the passive sign convention for ideal sources will therefore require that
we draw their $i - v$ characteristics in the positive voltage, negative current quadrant
of the axes as shown in Figure 1.6. Dependent sources simply require the additional specification of their output in terms of the control input. Now recall that for
passive components, in addition to defining the reference direction for current, we
also attached significance to the sense of the voltage change across a component
and indeed chose to work with the voltage drop. We now adopt the same convention for the various sources. The voltage drop is in the same sense as the *reference
direction* for current and a single arrow may be used, as for passive components, to

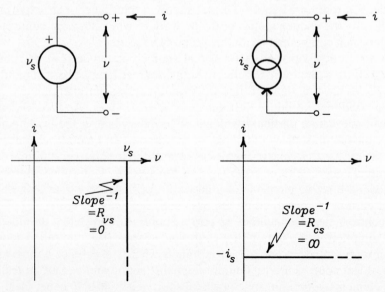

Figure 1.6

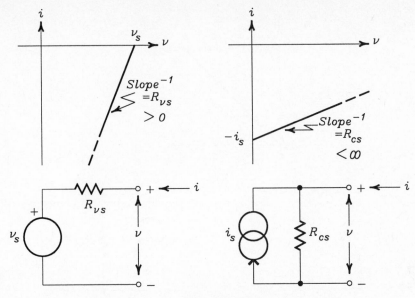

Figure 1.7

characterize the sense of both the voltage change and the current.

The current–voltage characteristic in each case specifies the properties of the source as perceived at its terminals and is therefore also called its *terminal characteristic*. The reciprocal of the slope of the current–voltage characteristic has the dimensions of resistance. It is the *internal resistance* or *source resistance* and will be seen to be zero for the ideal voltage source, infinite for the ideal current source.

It follows from the characteristics that we have assigned to the ideal sources that we must never speak of *shorting-out* an ideal voltage source nor of leaving an ideal current source on *open circuit*. In the former case an infinite current would flow since a voltage source will maintain its voltage across any resistance. In a similar way an infinite voltage would be maintained across an open circuit by a current source. (Often when the context is clear, as it is now, we may dispense with continued use of the word ideal.) To render such sources ineffectual we shall have to say that we *replace* a voltage source by a short circuit and *replace* a current source by an open circuit.

Practical sources, on the other hand, are real devices such as the batteries or laboratory bench power supply energizing a circuit or the transducers and active circuits providing inputs. They are properly called *non-ideal independent sources*. A practical voltage source will be less than ideal by virtue of its internal resistance being greater than zero. Conversely, a practical current source will have an internal resistance which is less than infinite. The terminal characteristics will therefore be as shown in Figure 1.7. In each case a representation in terms of a discrete ideal source and a discrete resistor is shown below the terminal characteristic. For the voltage source the voltage of the supposed internal ideal source is that which appears at the output terminals of the actual source when no current is flowing, that is, *the open-circuit voltage*. The internal resistance in series with the voltage source precludes the possibility of infinite current flowing when the real source is

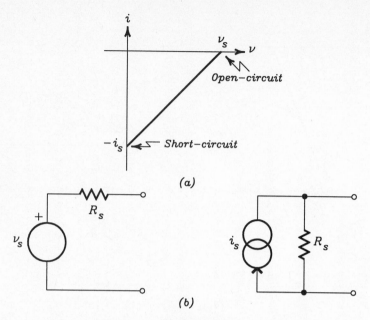

Figure 1.8

short-circuited. The whole of the source voltage then appears across the internal series resistor. For the current source, on the other hand, the current of the supposed internal ideal source is that which flows when the voltage across the terminals of the actual source is zero, that is, the *short-circuit current*. The shunt resistor on the current source ensures that the voltage does not become infinite when the real source is open-circuited. All of the current available from the current source is then taken by the internal shunt resistor. It must be emphasized at this point that these are only *models* of the real sources. A battery is an electrochemical process while a bench power unit is an assembly of many discrete components connected to the a.c. supply.

The inverse of the above argument is that, given a source of low internal resistance it would seem appropriate to describe it as a (practical) voltage source. Similarly, for a source of high internal resistance the most appropriate description would be as a (practical) current source. However, while these may each be the most obvious interpretation there is no inherent reason why we should not model the former as a current source and shunt resistor and the latter as a voltage source and series resistor. To illustrate this point consider the terminal characteristic, in Figure 1.8(a), of a source having an open-circuit voltage v_s and a short-circuit current i_s. The internal resistance, R_s, is the reciprocal of the slope, that is, $R_s = v_s/i_s$ and we are not concerned for the moment whether this is large or small. If only the open-circuit voltage were specified then the short-circuit current could be determined from $i_s = v_s/R_s$. Conversely, if only the short-circuit current were specified then the open-circuit voltage could be determined from $v_s = i_s R_s$. *Two* models of the source can therefore be given, Figure 1.8(b), *both* of which are consistent with the terminal characteristic. In one, the source resistance R_s is in series with an ideal source the output of which is the open-circuit voltage v_s. In the other, R_s is the shunt

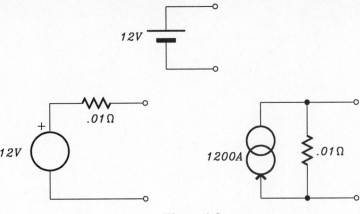

Figure 1.9

resistor on an ideal current source the output of which is the short-circuit current i_s.

We may decide at any time to replace one representation of a real source with the other. When we do so we say that we are performing a *source transformation*. While both of the representations connected by a source transformation are equally acceptable in describing the terminal characteristics, one will usually be the more physically reasonable.

Example 1.2 As an example consider an ordinary 12 V car battery which has an internal resistance of 0.01 Ω (Figure 1.9). The short-circuit current of the battery would be 1200 A. Therefore in addition to regarding the battery as an ideal independent voltage source of 12 V with a 0.01 Ω series resistor we could equally well consider it to be an ideal independent current source of 1200 A shunted by a 0.01 Ω resistor.

The idea of 1200 A continuously circulating within our battery is fundamentally less acceptable than the former model. In light-detecting devices, such as photodiodes or photomultiplier tubes, the number of electrons excited is in proportion to the light intensity. These are therefore good examples of cases where a description as a current source is appropriate. It may be noticed that when either of the representations of a practical source, related by a source transformation, is rendered dead in the way described earlier the same dead source, a resistor, remains.

While the source resistance determines the reciprocal of the slope of the terminal characteristic, the load resistance fixes the position of the operating point on the terminal characteristic (Figure 1.10). At the operating point v is the voltage across the load and i is the current *from* the most positive terminal so $v/i = -R_L^{-1}$ where R_L is the load resistance. We are still working in terms of a passive sign convention for the source so that i will be negative and the sign of the resistance positive as required. (On the terminal characteristic of the *load* the resistance is the reciprocal slope of the characteristic.) The operating point therefore lies at the intersection of a line with slope $-R_L^{-1}$ through the origin, called *the load line*, and the terminal characteristic.

In the use of the term 'practical source' we have recognized the non-ideality of sources. It may be useful at this stage to point out that most commercial

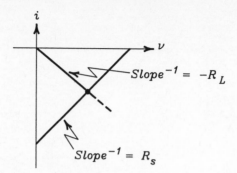

Figure 1.10

power supplies of any reasonable quality in fact come very close to being ideal but only over a limited range of operation. A typical characteristic is shown in Figure 1.11(a). This corresponds to a source capable of substantially constant voltage operation (typically to 0.01%) for a load resistance down to $R_L = |v_m/i_m|$. For any lower load resistance the maximum current capacity would be exceeded and the supply would adopt a constant current mode of operation. The transition occurs in the opposite sense as the load resistance is reduced from high values. The current setting therefore provides the current limit in the constant voltage mode and vice versa. The particular mode of operation will be determined not simply by the voltage and current settings but also by the load resistance. When the supply provides only 'current limiting' operation the regulation of current will be further from the ideal (Figure 1.11(b)).

When giving examples to illustrate the various theorems it is customary to assume that when a voltage source is required it may be taken to be ideal and similarly for a current source. This is not simply a labour-saving device but very much in accord with what would happen in practice. If we propose to provide a source of voltage to a circuit then we should certainly expect that its internal resistance is negligible compared with the circuit resistance. Conversely, we should expect the circuit resistance to be small compared with the source resistance for a current source. We shall adopt this practice in what follows.

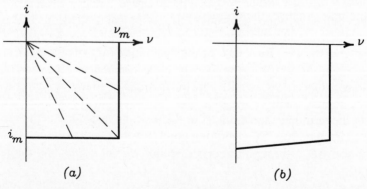

Figure 1.11

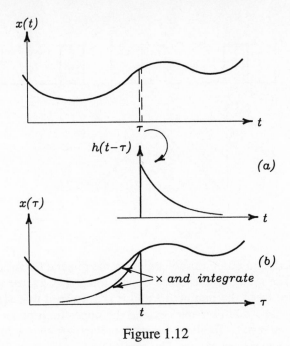

Figure 1.12

1.6   Stability and Causality

Finally among the introductory ideas we must make some other formal requirements of the networks with which we can deal. We require also that they are stable
and causal. By the first we mean that we expect the output to drop to zero when
the input is removed – it forgets the past. Secondly, we expect there is no output
until the input is applied – it cannot predict the future. A more quantitative statement may be made by considering how we envisage an output being obtained from
an input. Any input may be imagined as a series of very sharp spikes, impulses,
in infinitely rapid succession. Each such impulse can be described in terms of the
so-called *unit impulse* and an appropriate multiplier. The unit impulse is described
more fully in Chapter 6. The multiplier is the area of an elementary rectangle at
$t = \tau$ of width $\Delta\tau$, in the limit as $\Delta\tau \to 0$, and height $x(\tau)$ (Figure 1.12). The response to the unit impulse, the normalized impulse response, is usually regarded
as one of the most fundamental properties of a system. The output for an unit impulse at $t = 0$ is $h(t)$, say. For a unit impulse at $t = \tau$ one shifts the origin, writing
$h(t - \tau)$. The output for the input represented by the elementary rectangular area
$x(\tau)\Delta\tau$ is $x(\tau)\Delta\tau\, h(t - \tau)$ and the net output at time t due to the succession of all
such impulses that have occurred up to time t will be the sum of all such impulse
responses:

$$
\begin{aligned}
y(t) &= \lim_{\Delta\tau \to 0} \sum_0^t x(\tau)h(t - \tau)\Delta\tau \\
&= \int_0^t x(\tau)h(t - \tau)d\tau
\end{aligned}
$$

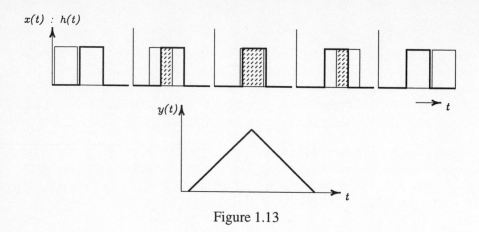

Figure 1.13

where τ serves as a dummy time variable to allow the integral for the output at time t to be obtained. The system represented by this integral is *causal* as only inputs up to t are contributing to the output at t and it is presumed that $x(t)$ itself starts at $t = 0$. For a causal system we may extend the upper limit of integration to ∞ without changing the result. To allow for inputs which are non-zero before $t = 0$ we may extend the lower limit to $-\infty$. The result:

$$y(t) = \int_{-\infty}^{\infty} x(\tau)\, h(t-\tau)d\tau$$

is known as *the convolution integral*. A system for which $y(t) \to$ a constant value as $x(t) \to 0$ is a *stable* system. It is often stated that the system should settle to zero, but so long as it settles to the same constant value this is sufficient.

The convolution integral is the basis for a very general approach to system analysis which is not to be the emphasis here. However, it will be useful in relation to topics which follow later to say a little more about it now. Note that in performing a convolution the integration is carried out over the (dummy) time variable τ, *not* the time t at which the output is required. The impulse response appears as a function of $h(\tau)$ in which τ is sign-reversed, $h(-\tau)$, and which is shifted by t to produce $h(t-\tau)$. A so-called graphical interpretation, Figure 1.12(b), can therefore be given to convolution in which we superimpose the impulse response, time-reversed and shifted to t, on the input function and find the area under the curve which is the product of the two functions, that is, integrate the product. The result yields the output at time t. In fact convolution can be defined between any two functions without them necessarily being connected to properties of an electronic system.

Example 1.3 A simple example is provided by the convolution of two rectangular pulses (Figure 1.13). As one function is swept through the other the area of overlap will first increase, then decrease, and this is reflected in the result. The output does not, in this case, bear any simple relationship to either of the convolved functions.

Summary

It has been emphasized that the role of an electronic engineer is as a *designer* rather than simply a constructor of circuits. Circuits in their turn should be seen as a variety of *physical system*. The particular systems chosen for study here are *linear* in that they are governed by the *principle of superposition*, that is, a sum of inputs produces an output which is the sum of the outputs produced by the inputs applied separately.

Passive and active systems have been distinguished, the former characterized by a net dissipation of power. The simple components resistor, capacitor and inductor were identified by their current–voltage characteristics and a *passive sign convention* was adopted (current *into* the most positive terminal is considered to be in the positive direction).

The passive sign convention was also adopted for the power supplies used to energize electronic systems and for the transducers used to excite them, both described as *sources*, for which the current–voltage characteristic, known here as the *terminal characteristic*, is dependent upon the circuits to which they are connected. *Ideal independent sources* were defined as a means of modelling *non-ideal* (or *practical*) *sources*. The realization of any particular terminal characteristic was seen to be possible using either an ideal independent voltage source with a series resistor, or an ideal independent current source with a shunt resistor so leading to the definition of a *source transformation* from one representation to the other. Should it be necessary to suppress an ideal source (render it *dead*) we must be careful to say that we *replace* an ideal independent voltage source with a short circuit and *replace* an ideal independent current source with an open circuit. It produces absurd results to try to short-circuit the former or to open-circuit the latter. It is reassuring that the dead source produced by suppressing the ideal source used to model a practical source is simply the source resistance whether we choose the voltage source and series resistor representation or the current source and shunt resistor.

Ideal dependent (or *controlled*) *sources* where the output is dependent upon a current or voltage elsewhere in the system were also defined. Such sources are required in the modelling of active circuits.

Stability and *causality* were seen as formal requirements on the properties of systems in that we expect the output to drop to zero when the input is removed (the system forgets the past) and no output should appear before the input (the system cannot predict the future). The latter was also discussed in connection with the *convolution* of the input with the impulse response of a system as a means of visualizing the output.

Problems

1.1 Determine whether or not the systems characterized by the following input–output relations are linear:

    (a) $y = x^2$

    (b) $y = t(dx/dt)$

    (c) $y = x(dx/dt)$

1.2 Give *both* circuit models (voltage source plus resistor and current source plus resistor) corresponding to the terminal characteristic of Figure 1.14.

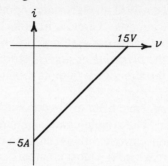

Figure 1.14

1.3 Perform source transformations on each of the circuit models of Figure 1.15.

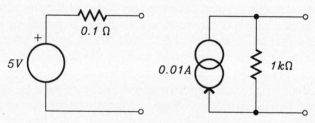

Figure 1.15

1.4 Using a series of sketches deduce the effect of convolving a function e^{-t} (for future reference this is the impulse response of a resistor–capacitor circuit with a 1 second time constant) with a unit amplitude sine wave of frequency 0.01 Hz. Examine the effect of increasing the frequency to 1 Hz.

2

Network Equations

2.1 Kirchhoff's Laws

As stated previously, electronic circuits are examples of physical systems and as such are bound by the fundamental physical laws. In particular, at the energies at which circuits are typically employed the conservation both of energy and of charge is observed. Consequently, if a charge q is carried around the simple circuit shown in Figure 2.1(a) the energy imparted by the battery is dissipated, as joule heat, in the resistor. That is $qv_B = qv_R$ or $v_B - v_R = 0$. For the junction of conductors (node) shown in Figure 2.1(b) we must have that $i_1 + i_2 - i_3 = 0$ if charge is not to be spontaneously created or annihilated at the node, that is, if it is conserved. The conservation laws for circuits are generalized in two laws:

Kirchhoff's voltage law (KVL): at any instant the algebraic sum of the voltages around any closed path (or loop) within a network is zero.

$$\sum_n v_n(t) = 0 \tag{2.1}$$

Kirchhoff's current law (KCL): at any instant the algebraic sum of currents leaving any junction (or node) in a network is zero.

$$\sum_n i_n(t) = 0 \tag{2.2}$$

The *algebraic sum* means that we first express each sum symbolically, adopting sign conventions for the polarity of voltages around a loop and for the direction of currents at a node. There are a number of possibilities in the choice of such

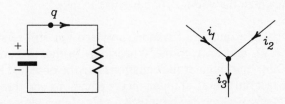

Figure 2.1

15

KVL:

$$v_{R_1} + v_{R_2} - v_s = 0$$
$$iR_1 + iR_2 - v_s = 0$$
$$v_s = iR_1 + iR_2$$

Total resistance:
$$R = \frac{v_s}{i}$$
$$= R_1 + R_2$$

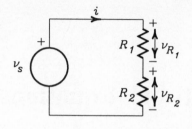

KCL:

$$i_{R_1} + i_{R_2} - i_s = 0$$
$$\frac{v}{R_1} + \frac{v}{R_2} - i_s = 0$$
$$i_s = \frac{v}{R_1} + \frac{v}{R_2}$$
$$\text{and } \frac{i_s}{v} = \frac{1}{R}$$
$$= \frac{1}{R_1} + \frac{1}{R_2}$$

where R is the total resistance.

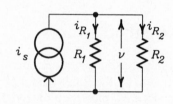

Figure 2.2

conventions. We decided in Chapter 1 to attach significance to the voltage drop across a component or across a source and noted that the voltage drops are in the same sense as the reference direction for current in a component or a source. We now choose to sum the voltage drops in a clockwise sense around a loop. A voltage which decreases with a clockwise traversal is therefore taken as positive, otherwise it is negative. For currents we take those directed away from a node to be positive. We subsequently assign the appropriate numerical value to each. Requiring that the sum be taken *at any instant* means that, besides describing the d.c. case, the laws need no reformulation to deal with time-varying voltages and currents. We shall try to adhere to the use of the word *network* to describe interconnected components, rather than *circuit*, as the latter also has the connotation of a closed path. However, established usage will dictate that we occasionally depart from rule as in the case of 'equivalent circuit'.

Example 2.1 We first illustrate the application of equations (2.1) and (2.2) by reference to two trivial examples with which the reader is almost certainly already familiar: resistors in series and in parallel (Figure 2.2). The subscripts indicate the network branches in which the voltages or currents are taken.

Example 2.2 Now as an example of a more involved but still relatively simple network, we apply KVL and KCL to the Wheatstone bridge (Figure 2.3). Upon inspection it is apparent that some of the equations may be obtained by combining others. There are in fact only three unique voltage equations and three unique current equations, a total of six which can be solved for six of the unknown branch voltages and currents. The six current–voltage relationships for the individual branches

$$v_{R_1} + v_m + v_{R_4} - v_s = 0$$
$$v_{R_2} - v_m + v_{R_3} - v_s = 0$$
$$v_{R_1} + v_{R_3} - v_s = 0$$
$$v_{R_2} + v_{R_4} - v_s = 0$$
$$v_{R_1} + v_m - v_{R_2} = 0$$
$$v_{R_3} - v_{R_4} - v_m = 0$$
$$v_{R_1} + v_{R_3} - v_{R_4} - v_{R_2} = 0$$
$$i_{R_1} + i_{R_2} - i = 0$$
$$i_{R_3} + i_m - i_{R_1} = 0$$
$$i_{R_4} - i_m - i_{R_2} = 0$$
$$i - i_{R_3} - i_{R_4} = 0$$

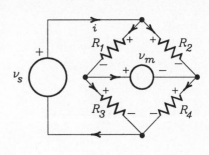

Figure 2.3

allow the remaining six currents and voltages to be determined.

While it is fairly straightforward to see how to reduce the number of equations in this case, methods for deciding upon the minimum set of equations in a general case must be obtained by resort to *graph theory*. A *graph* corresponding to a particular network is produced by replacing each passive component and each source by a simple line. Only in the sense that a line drawing is produced is there any connection with a graph of the *f(x) versus x* sort. Central to graph theory is the concept of the *tree*, which is the minimum set of edges (or, in this context, *branches*) connecting *all* of the nodes *without introducing any circuits*. For a network having B branches and N nodes any particular tree will have $(N-1)$ branches. (One further branch would complete a circuit.) The part of the network which is not included in the tree is the *cotree*. It follows that the cotree will have $(B-(N-1))$ edges. Each edge of the cotree is a link which will complete a fundamental (or f-) circuit when added to the tree. The f-circuits provide a systematic way of drafting the minimum set of independent KVL equations but an equal number of circuit equations chosen in some other way may also constitute a basis set. In whichever way it is chosen, the basis set must contain $B-N+1$ equations. Examples of trees, cotrees and f-circuits for the Wheatstone bridge are shown in Figure 2.4(a).

Graph theory also discusses what may be learned by enumerating the various ways that a complete network can be cut into two separate components or sets of nodes. The cutsets, as they are called, are sets of branches, which, if cut, produce such separated components. The cutsets for the Wheatstone bridge are shown in Figure 2.4(b). For each set of branches incident at a cut we shall expect charge to be conserved. Generalizing KCL to include all the cutsets for the Wheatstone bridge we generate a total of six equations, two more than for the nodes alone. Specifying that a cutset should contain only one branch of any chosen tree defines the number of fundamental (or f-) cutsets. For a network of B branches and N nodes there are $N-1$ tree branches and that is the required minimum number of equations. The f-cutsets provide a systematic way to write the independent set of KCL equations. An equal number of equations chosen in some other way may also constitute a basis set and, in particular, there is no reason why they should not be written for $N-1$ of the nodes. The important results of graph theory are embodied in a set of so-called

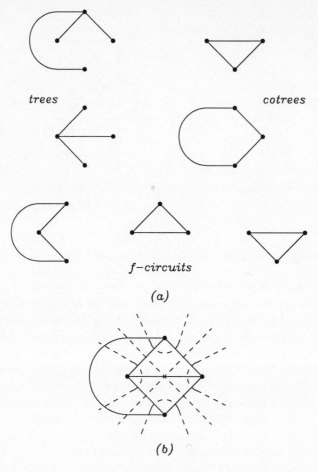

trees   *cotrees*

f−circuits

(a)

(b)

Figure 2.4

secondary equations which we now examine.

2.2 The Node and Loop Equations

From our brief review of graph theory we see that for a network of B branches and N nodes we can always write $B - N + 1$ independent KVL equations and $N - 1$ independent KCL equations. We have a total of B equations and can therefore always solve for B of the unknown branch voltages and currents, the remaining voltages and currents being determined by the current–voltage relationships for the branches. In fact the minimum requirement for a complete specification of a network is *either* $B - N + 1$ KVL equations *or* $N - 1$ KCL equations, together with the B branch current–voltage relations. That this is so may be shown most succinctly using matrix algebra, another powerful tool for the analysis of circuits.

Example 2.3 The application of matrix algebra may be illustrated by again taking as an example the Wheatstone bridge, represented now by its directed graph (Figure

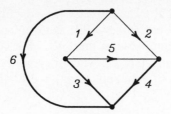

Figure 2.5

2.5). The heavy lines in the figure define a particular tree chosen to identify $N-1$ nodes by cutting its branches while the lighter lines are the corresponding cotree the branches of which each complete an f-circuit. In the directed graph a sense is assigned to each branch (or edge) corresponding to the reference direction for current (which is the same as the sense of the voltage drop) in the branch. We use 0 and ±1 to indicate whether a given voltage is present in a particular closed loop and whether the sense of the branch is with $(+)$, or counter $(-)$, to the direction of a clockwise traversal of a loop, 0 and ±1 also show which currents are directed away from $(+)$, or towards $(-)$, a given node. Then, with the numbering of the branches as shown in Figure 2.5, a basis set of KVL and KCL equations for the Wheatstone bridge may be written in matrix form as follows:

$$
\begin{bmatrix}
1 & 0 & 1 & 0 & 0 & -1 \\
0 & 1 & 0 & 1 & 0 & -1 \\
0 & 0 & -1 & 1 & 1 & 0
\end{bmatrix}
\begin{bmatrix}
v_1 \\ v_2 \\ v_3 \\ v_4 \\ v_5 \\ v_6
\end{bmatrix}
=
\begin{bmatrix}
0 \\ 0 \\ 0
\end{bmatrix}
\tag{2.3a}
$$

$$
\begin{bmatrix}
1 & 1 & 0 & 0 & 0 & 1 \\
-1 & 0 & 1 & 0 & 1 & 0 \\
0 & -1 & 0 & 1 & -1 & 0
\end{bmatrix}
\begin{bmatrix}
i_1 \\ i_2 \\ i_3 \\ i_4 \\ i_5 \\ i_6
\end{bmatrix}
=
\begin{bmatrix}
0 \\ 0 \\ 0
\end{bmatrix}
\tag{2.3b}
$$

The result in the example may be expressed more compactly thus:

$$\mathbf{B}\mathbf{v}(t) = 0 \tag{2.4a}$$
$$\mathbf{Q}\mathbf{i}(t) = 0 \tag{2.4b}$$

where $\mathbf{B}$, of order $(B-N+1) \times B$, and $\mathbf{Q}$, of order $(N-1) \times B$, are so-called basis circuit and basis cutset matrices. In fact, because in this case the KCL equations have been written for $N-1$ of the nodes, the matrix $\mathbf{B}$ is identical to what is called the basis incidence matrix. An incidence matrix refers just to currents incident at a node and will be a sub-matrix of the complete cutset matrix. A very interesting property of $\mathbf{B}$ and $\mathbf{Q}$ is that, if their columns are arranged in the same branch order, then

$$\mathbf{Q}\mathbf{B}^T = 0 \tag{2.5}$$

where $\mathbf{B}^T$ is the transpose of $\mathbf{B}$. The property may readily be established for the matrices appropriate to the Wheatstone bridge as given above. Equation (2.5) describes the intimate connection between the KVL and KCL equations. It is not surprising that such a relationship exists when they are both describing the same network. A general solution of equations (2.4a) and (2.4b) may be given as follows:

$$\mathbf{v}(t) \;=\; \mathbf{Q}^T \mathbf{v}_r(t) \tag{2.6a}$$
$$\mathbf{i}(t) \;=\; \mathbf{B}^T \mathbf{i}_m(t) \tag{2.6b}$$

To show that this is so we pre-multiply equation (2.6a) by $\mathbf{B}$ and equation (2.6b) by $\mathbf{Q}$ and then use equation (2.5), or its transpose, as appropriate. The elements of the $(N-1)$ component voltage vector, $\mathbf{v}_r(t)$, may be identified as voltage differences around the network. The exact choice from among those available is arbitrary but a convention is usually adopted. In a similar way the $(B - N + 1)$ component current vector, $\mathbf{i}_m(t)$, may be identified as currents circulating in the various circuits within the network which combine to form the branch currents $\mathbf{i}(t)$. Again the exact choice from among those available is arbitrary but a convention is adopted.

The two approaches which exploit the possibility of such minimal sets of equations are the node equations (or node-voltage) method and the loop equations (or loop-current) method. For networks which are *planar*, that is, may be drawn on a flat surface without any crossing lines, the latter becomes the mesh equations (or mesh-current) method. In the node equations emphasis is given to determining the *voltages* around the network relating these via *KCL equations* at a selection of the nodes. In the loop equations emphasis is given to determining the *currents* in the network relating these via *KVL equations* for loops within the network. While matrix algebra may be used to give complete justification to the methods it will be seen for the most part that the results are almost self-evident.

The node equations are written by recognizing that the N node voltages alone can be used to define the B branch voltages. In fact, since branch voltages are node voltage differences, the voltages of $N-1$ nodes may be given with reference to one chosen node and its voltage set arbitrarily to zero. The branch currents may be obtained from the branch voltages using the branch current–voltage relationships and the KCL equations written for each of the $N-1$ nodes. The node equations may now be solved for the unknowns, the $N-1$ node voltages. One obtains the voltages and currents in the branches by retracing the steps leading to the node equations. If only two branches join at a particular node we have the option, in a first round of analysis, of treating those two branches as a single branch so reducing the number of equations by one for each such pair of branches. The voltages and currents for the component branches may be deduced at a later stage using the branch current–voltage relationships, the potential divider principle or most simply, if ideal sources are involved, by inspection.

To write the loop equations we introduce fictitious currents circulating in each of $B - N + 1$ network loops so that the current in any branch is the sum of the currents in the loops having that branch in common. KVL equations are then written for each loop, the voltage in each element being given by the branch current and the branch current–voltage relationship. The loop equations may be solved for the $B - N + 1$ loop currents. Again one retraces some of the earlier steps to obtain the

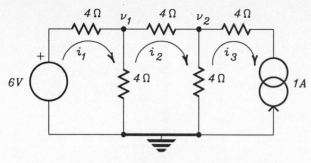

Figure 2.6

individual branch voltages and currents. The network loops for which the basis set KVL equations should be written are most easily identified when the network in question is planar. Then, regarding the network as a mesh screen, the perimeter of each 'hole' in the mesh is an f-circuit. We shall not pursue further the case of non-planar circuits.

Example 2.4 The node-voltage and mesh-current methods may be illustrated by application to the network shown in Figure 2.6. There are five nodes and seven network branches but we may reduce each of these numbers by two by regarding two pairs of branches each as one. Two node and three mesh equations will be required. (Notice that counting branch pairs as one does not affect the number of mesh equations required.) The node equations are:

$$\frac{v_1 - 6}{4} + \frac{v_1}{4} + \frac{v_1 - v_2}{4} = 0$$

$$\frac{v_2}{4} - 1 - \frac{v_1 - v_2}{4} = 0$$

Hence $v_1 = \frac{16}{5}$ V and $v_2 = \frac{18}{5}$ V

The mesh equations are:

$$4i_1 + 4(i_1 - i_2) - 6 = 0$$
$$4i_2 + 4(i_2 - i_3) - 4(i_1 - i_2) = 0$$
$$i_3 = -1$$

Hence $i_1 = \frac{7}{10}$ A and $i_2 = \frac{-1}{10}$ A

The mesh currents will be seen to be consistent with the node voltages obtained earlier.

An ideal voltage source that is the sole element in that branch has the effect of determining the branch voltage. The number of unknown branch voltages is reduced by one as is the number of unknown node voltages. The branch currents all remain to be determined. When the source is not alone in the branch the number of independent equations is not affected. In the case of a current source the branch current

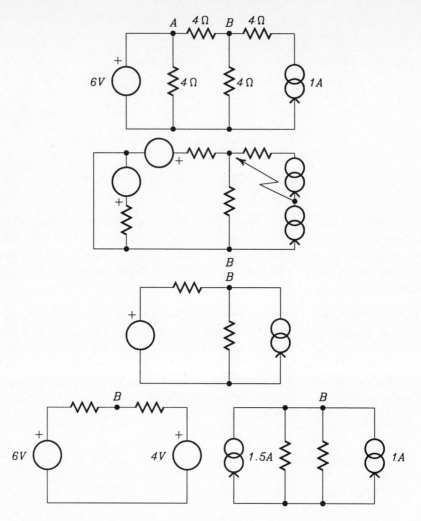

Figure 2.7

is determined whether or not the source is the sole element. The number of mesh currents is reduced by one but the number of branch voltages remains unchanged. The effect is demonstrated for the application of the mesh current method to the network given in Figure 2.6 where it will be seen that one mesh current is equated to the source current.

For simple networks the way to proceed is usually clear. In more complex situations it may be helpful to perform source transformations so that only current sources remain for node analysis and only voltage sources for mesh analysis. In preparation for doing this, voltage sources that are the sole element in a branch and current sources in series with other elements are shifted to equivalent positions in the network. One recognizes that a voltage source affects the voltage of every branch connected to it at a common node. The same effect would be achieved by providing an identical source in each of those branches instead of the original source. In a similar way a current source is seen as delivering current to one

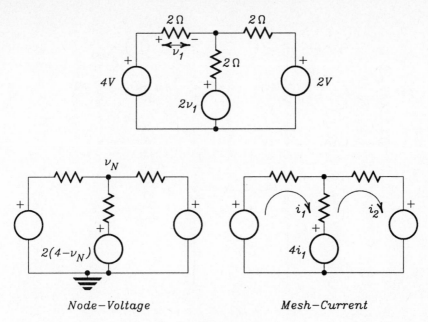

Figure 2.8

terminal while draining current from another. It may therefore be replaced by *two* sources the common terminal of which can be connected to any node in the network as it makes no net difference to that node. The effect of such source-shifting operations followed by source transformation is illustrated in Figure 2.7. The voltage source may be translated through node A at the same time becoming two sources but the empty branch left behind (a short circuit) now renders one of the two new sources ineffectual. The remaining voltage source is then transformed to its equivalent current source. Conversely, the original current source may be replaced by two sources and the common connection taken to any convenient node. The node may be chosen in this case so that only one of the two sources is effectual. Source transformation will then yield a voltage source. Node or mesh analysis in this case would be trivial as there are only two nodes and one mesh, however it does illustrate the procedure. Finally the role of dependent (or controlled) sources must be mentioned. When the node or mesh equations are written, any dependent source should be expressed in terms of the node voltages or mesh currents as appropriate making use as required of the branch current–voltage relationships. An example is given in Figure 2.8.

2.3 Tellegen's Theorem

Our main purpose in invoking graph theory was to reduce to a minimum the number of independent KVL and KCL equations required to characterize a network. However, it may have become apparent to the reader that the KVL and KCL equations as expressed in equations (2.3) and (2.4) actually characterize the graph representing the circuit rather than the circuit itself. The properties of the circuit elements in the

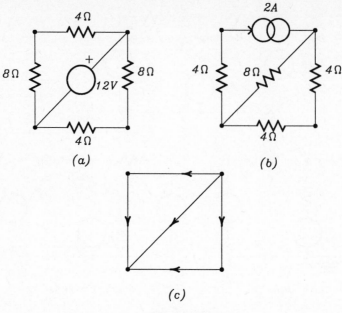

(c)

Figure 2.9

various branches are of no immediate importance in those equations. Furthermore, the KVL and KCL equations each have only one solution regardless of the actual circuit they might represent. We may illustrate this with respect to the two circuits in Figure 2.9 which clearly are physically quite different but which may both be respresented by the same directed graph, Figure 2.9(c). We may write the KVL and KCL equations for the networks in Figure 2.9(a) in the form of equation (2.4):

$$\mathbf{B}\mathbf{v}(t) = 0$$
$$\mathbf{Q}\mathbf{i}(t) = 0$$

knowing that they have the solution given in equation (2.6):

$$\mathbf{v}(t) = \mathbf{Q}^T\mathbf{v}_r(t)$$
$$\mathbf{i}(t) = \mathbf{B}^T\mathbf{i}_m(t)$$

Similarly we may write equations identical in form for the network of Figure 2.9(b) except that we now use upper-case symbols to distinguish this set of equations from the first:

$$\mathbf{B}\mathbf{V}(t) = 0$$
$$\mathbf{Q}\mathbf{I}(t) = 0$$

with solution:

$$\mathbf{V}(t) = \mathbf{Q}^T\mathbf{V}_r(t)$$
$$\mathbf{I}(t) = \mathbf{B}^T\mathbf{I}_m(t)$$

The two sets of solutions are, as we have discussed above, independent of the circuit elements and related only to the common graph. The two sets of solutions are therefore the same. It follows that we may take products between voltages for one network and currents for the other. Using the solutions given above, equation (2.5) and the result that if:

$$\mathbf{v}(t) = \mathbf{Q}^T \mathbf{v}_r(t)$$

then

$$\mathbf{v}^T(t) = \{\mathbf{Q}^T \mathbf{v}_r(t)\}^T = \mathbf{v}_r^T(t)\mathbf{Q}$$

and if:

$$\mathbf{i}(t) = \mathbf{B}^T \mathbf{i}_r(t)$$

then

$$\mathbf{i}^T(t) = \{\mathbf{B}^T \mathbf{i}_r(t)\}^T = \mathbf{i}_r^T(t)\mathbf{B}$$

We find:

$$\mathbf{v}^T(t)\mathbf{I}(t) = \mathbf{v}_r^T(t)\mathbf{Q}\mathbf{B}^T\mathbf{I}_r(t) = 0 \qquad (2.7a)$$

Similarly:

$$\mathbf{V}^T(t)\mathbf{i}(t) = 0 \qquad (2.7b)$$
$$\mathbf{i}^T(t)\mathbf{V}(t) = 0 \qquad (2.7c)$$
$$\mathbf{I}^T(t)\mathbf{v}(t) = 0 \qquad (2.7d)$$

Any one of these equations, or all collectively, may be taken to be a statement of Tellegen's theorem.

Example 2.5 For the networks in Figure 2.9 it may be shown, almost by inspection, that:

$$\mathbf{v} = \begin{bmatrix} 4 \\ 8 \\ 4 \\ 8 \\ 12 \end{bmatrix}, \mathbf{i} = \begin{bmatrix} 1 \\ 1 \\ 1 \\ 1 \\ -2 \end{bmatrix}, \mathbf{V} = \begin{bmatrix} 16 \\ 4 \\ 4 \\ -8 \\ 8 \end{bmatrix}, \mathbf{I} = \begin{bmatrix} -2 \\ 1 \\ 1 \\ -2 \\ 1 \end{bmatrix}$$

Hence:

$$\mathbf{v}^T(t)\mathbf{I}(t) = \begin{bmatrix} 4 & 8 & 4 & 8 & 12 \end{bmatrix} \begin{bmatrix} -2 \\ 1 \\ 1 \\ -2 \\ 1 \end{bmatrix}$$

$$= -8 + 8 + 4 - 16 + 12 = 0$$

and

$$\mathbf{V}^T(t)\mathbf{i}(t) = \begin{bmatrix} 16 & 4 & 4 & -8 & 8 \end{bmatrix} \begin{bmatrix} 1 \\ 1 \\ 1 \\ 1 \\ -2 \end{bmatrix}$$

$$= 16 + 4 + 4 - 8 - 16 = 0$$

The sum of the products of voltages and currents for the branches of any one network, expressed as the matrix products $\mathbf{v}^T(t)\mathbf{i}(t)$ or $\mathbf{V}^T(t)\mathbf{I}(t)$, may also be shown to be zero by the same procedure as for equations (2.7). The result shows that there is no net power dissipation in the network, that is, the energy delivered by the sources at each instant is consumed, possibly to be stored, by the components. For the last statement to remain true we require the simultaneous conservation of both energy and charge, as expressed separately in Kirchhoff's laws. Tellegen's theorem may therefore be interpreted as a joint statement of KVL and KCL for a single network. While emphasis is sometimes given to the formulation of Tellegen's theorem in this form, that is, as it applies to power conservation in a single network, this will be seen to be almost trivial compared with the light that the theorem sheds on the joint solution of networks. It can be of great assistance in implementing and checking computer-based techniques for circuit analysis.

2.4   The State Equations

We have confined ourselves so far to a consideration of purely resistive networks but there is no reason why the passive components should not in fact include capacitors and inductors in addition to resistors. When we apply Kirchhoff's laws or node and mesh analysis to such networks we obtain a set of first-order integro-differential equations. As an example consider a voltage source connected to an inductor, a capacitor and a resistor in series (an LCR circuit) as shown in Figure 2.10. The voltage across the inductor is given by Lenz's law ($v_L = L\,di/dt$) while the capacitor voltage is related to the charge which is itself found by integrating the current:

$$v_o = \frac{q}{C} = \frac{1}{C}\int i\,dt \tag{2.8}$$

An application of Kirchhoff's voltage law yields the following:

$$L\frac{di}{dt} + Ri + \frac{1}{C}\int i\,dt = v_s \tag{2.9}$$

Differentiating equation (2.8) twice to obtain i and di/dt and substituting in equation (2.9) we obtain the second-order differential equation:

$$LC\frac{d^2v_o}{dt^2} + RC\frac{dv_o}{dt} + v_o = v_s \tag{2.10}$$

This is the *system differential equation* connecting the input and the output. Much of the remaining subject matter of this and similar texts is concerned with how to

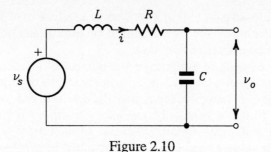

Figure 2.10

avoid having to solve such equations and indeed even having to write them down in the first place.

An alternative to producing the system equation is to characterize the network in terms of a set of first-order equations which describe the time development of certain parameters chosen to represent *the state* of the system. By choosing the current through the inductor and the voltage across the capacitor as state variables, appropriate equations are obtained, one from equation (2.9) by introducing the capacitor voltage explicitly and rearranging, the other by differentiating equation (2.8):

$$\frac{di}{dt} = \frac{-R}{L}i - \frac{1}{L}v_o + \frac{1}{L}v_s \qquad (2.11a)$$

$$\frac{dv_o}{dt} = \frac{1}{C}i \qquad (2.11b)$$

Equations (2.11) are jointly known as the *state equations*. One may also develop the state equation from the system equation. It then becomes clear that the number of first-order equations required is determined by the order of the system as reflected in the order of the system equation. However, the 'state' so defined may no longer have a one-to-one connection to network currents and voltages. This is another area where matrix algebra is of great value, allowing a very compact representation of the state equation for any order of system. In the above case we could write:

$$\begin{bmatrix} \dfrac{di}{dt} \\ \dfrac{dv_o}{dt} \end{bmatrix} = \begin{bmatrix} \dfrac{-R}{L} & \dfrac{-1}{L} \\ \dfrac{1}{C} & 0 \end{bmatrix} \begin{bmatrix} i \\ v_o \end{bmatrix} + \begin{bmatrix} \dfrac{1}{L} \\ 0 \end{bmatrix} \begin{bmatrix} v_s \\ 0 \end{bmatrix}$$

We shall just mention that in this formalism the state variables are regarded as components of a vector, the *state vector*. The state equation, in fact, often gives more information about the system than is really required. Its great merit is that it is very easily adapted to deal with discrete time (digital) systems, to time-varying systems and to non-linear systems.

Summary

Once again discussing electronic circuits in the context of their properties as physical systems we saw that the conservation laws for energy and charge take the special form known as *Kirchhoff's laws*. Kirchhoff's voltage law governing the sum of voltages for any closed path within a network is a statement of the conservation of energy, while Kirchhoff's law for the sum of currents at any node in a network is a statement of the conservation of charge.

As a means of reducing to a minimum the number of equations produced by an application of Kirchhoff's laws we drew upon the results of *graph theory* which deals with the *topology* of networks rather than the details of the components in the network branches. In this way a set of *secondary equations* is produced, known as the *node-voltage* and *mesh-current* equations. In the former a solution for the network (the branch voltages and currents) is obtained by writing Kirchhoff current law sums at all but one of the nodes in terms of the voltages of those nodes each referred to the remaining node. In the latter case Kirchhoff voltage law sums are written for each mesh in the network in terms of fictitious currents said to be circulating in each mesh (the mesh currents).

The overriding importance of the topology of networks was demonstrated in an example where the branch components were very different but the network topology was identical. An intimate connection between the branch voltages for one network and the branch currents for the other could be seen. The result is stated as *Tellegen's theorem* for all networks.

The idea of the *state* of a system was introduced by reference to a simple LCR network. A straightforward way of obtaining the *state variables* is to choose the current through each inductor and the voltage across each capacitor. The variables are then related by a series of coupled first-order differential equations, the *state equations*. The state variables may also be developed from the system differential equation when their physical significance may not be so clear. The variables for a system are said to be the components of a *state vector*. The state vector is a useful concept in making the transition to non-linear systems, time-dependent systems and digital systems.

Problems

2.1 Show by the application of Kirchhoff's laws that, for the potential divider shown in Figure 2.11, $v_o = \dfrac{R_2}{R_1 + R_2} v_i$

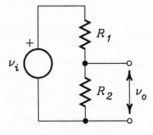

Figure 2.11

2.2 By the application of Kirchhoff's laws to the circuits of Figure 2.12 find v_1 in each case ($2v_1$ is a controlled source).

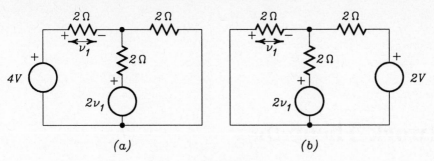

Figure 2.12

2.3 Show that, for the incidence matrix (**Q**) and the circuit matrix (**B**) of the Wheatstone bridge:

$$\mathbf{Q}\mathbf{B}^T = 0$$

2.4 Perform both node and mesh analysis for the circuit shown in Figure 2.13 to determine all branch voltages and currents.

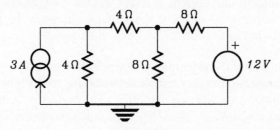

Figure 2.13

2.5 Complete the node and mesh analysis for the network of Figure 2.8.

3

Network Theorems

3.1 Superposition in Network Analysis

The principle of superposition underpins the definition of linearity. Apart from this formal role, the principle of superposition has great practical importance in analyzing networks where more than one source is present. It may be applied in the very many cases where we do not seek a complete solution, that is, detailed knowledge of the voltage and current in every branch as provided by node and mesh analysis, but rather need to know the voltage difference between just two points within the network or the current between those points.

Example 3.1 The utility of the principle of superposition in such applications may be illustrated by reference to the very simple network shown in Figure 3.1(a) which includes just two sources, in this case batteries. To obtain the voltage difference between A and B we could write down two KVL equations and solve for V_{AB}. Instead we can apply the prescription: *the net effect of the various sources is the sum of the effects of the sources applied separately, the others being temporarily sup-*

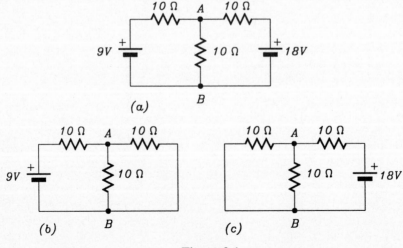

Figure 3.1

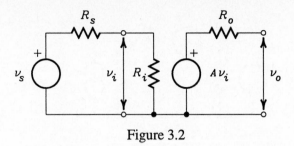

Figure 3.2

pressed. By suppressed we mean rendered ineffectual or dead in the way described in section 1.5. (*Replace* a voltage source with a short circuit, *replace* a current source with an open circuit.) In the above network the voltage V_{AB} may therefore be obtained as the sum of the voltages developed in the two circuits (b) and (c) in Figure 3.1. In each, one source has been suppressed. The combination of resistors in parallel is required in each case followed by the treatment of each network as a potential divider. The result is 3 V for (b) and 6 V for (c) giving a total of 9 V for V_{AB}.

When there are both independent voltage and current sources present in a network it is often useful to perform source transformations so that only voltage sources remain, in a case where a net voltage is required, or so that only current sources remain, where a net current is required. When a network includes a dependent (controlled) source then superposition must be applied with some refinement. To illustrate the need for this consider the network in Figure 3.2. It is the simplest representation of how a linear amplifier appears at its terminals to other networks connected to it. It is an *equivalent circuit*. (There will be more about equivalent circuits in section 3.2.) The circuit includes a dependent source to show how the output is driven by the input. Now let us take the contribution of each source in turn, suppressing other sources, with a view to calculating the output of the amplifier:

(i) suppress v_s: $v_i = 0$ therefore $Av_i = 0$ and $v_o = 0$

(ii) suppress Av_i: $v_o = 0$

The resultant is 0 and we have therefore completely killed the amplifier. The mistake has been to suppress the dependent source. This must be left to be suppressed by suppressing the source which controls it. We should omit step (ii) above. The output of the amplifier is therefore always Av_i and superposition really has no role to play in this instance. The correct procedure in such cases may be seen by considering an example.

Example 3.2 For the network in Figure 3.3(a) we wish to find v_0. We shall suppress the two independent sources in turn, one at a time. First suppress the current source (replace it by an open circuit), Figure 3.3(b). The current through the leftmost resistor is therefore 0.5 A and so $6i_1$ is 3 V. A further formal application of the principle of superposition could now be made but it is fairly evident that we have a 3 V source opposing a 5 V source with the resultant being 2 V. When we suppress

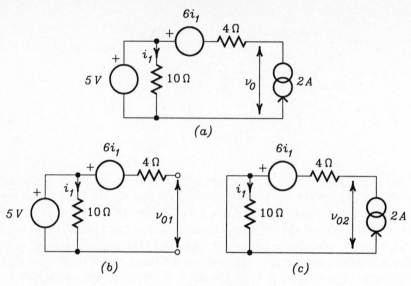

Figure 3.3

the independent voltage source (replace it by a short circuit), Figure 3.3(c), the left-most resistor is in parallel with the short circuit and so i_1 is zero. It follows that the dependent source is dead. A dead voltage source is characterized by its internal resistance which for an ideal source is zero. The contribution to v_0 is 8 V and the total is therefore 10 V.

3.2 Thevenin's and Norton's Theorems

We have just seen how the voltage difference between two points within a network or the current between those points may be obtained by an application of the principle of superposition. Often the points in question are the ends of a single branch. Even more frequently the branch of interest is either that which includes the load or that which includes an input device. In the first case the ends of the branch are output terminals and in the second case they are input terminals (Figures 3.4(a) and 3.4(b)). Given the linearity of the systems with which we are dealing we might expect to be able to partition the total network about the points between which we require the voltage difference or the current (Figures 3.4(c) and 3.4(d)), apply superposition to each of the two halves and then superpose the results. Depending on the final result required, a voltage or a current, the open-circuit voltage or the short-circuit current of each half would be obtained as a first step. To superpose these results in the second stage it would be necessary to treat each partial network as a simple source and resistor. The source voltage or current would be that calculated in the first stage. Such a reduced representation, or *equivalent circuit*, is therefore implicit in our understanding of how to apply superposition. Frequently, an equivalent circuit need only be developed for half the total network as the other half is a single branch, as mentioned above. It is an extremely useful concept showing what value of resistance a network will present to a source connected at its input termi-

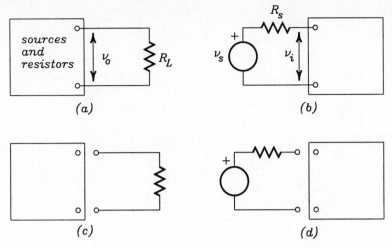

Figure 3.4

nals and whether it will perform more as a voltage source (low internal resistance) or as a current source (high internal resistance) when connections are made at its output terminals. Leaving aside the reasoning above, and the proofs to follow, it is largely a matter of common sense that such equivalent circuits should be available. We simply have to imagine the partial network in a black box which prohibits access to it except at its output terminals. As far as we know the box *might* contain just one resistor and one source. It is not possible to devise any test which could be performed at the terminals to identify the contents of the box. The reduction of a network to a single resistor and a single voltage source is known as Thevenin's theorem. The reduction to a single current source and a single resistor is Norton's theorem. Each as we shall see provides a template for the application of the principle of superposition. While, as just indicated, either may be obtained largely as a matter of common sense, we gain great insight into their application by proving them, which we do for Thevenin's theorem as follows.

First we partition the total network into the section for which we require the equivalent circuit and the section external to this, for which we wish to find the voltage difference. The external network may be any combination of components including sources but for the moment it is taken to be a simple resistor. The restriction will be removed later. The current through the external network is, say, i_1 (Figure 3.5(a)). We need not enquire how the driving network develops the current, suffice to say that it is available at its terminals. A variable ideal voltage source is now introduced as part of the external network in series with the resistor (Figure 3.5(b)) and adjusted until the current is reduced to zero. The driving network is delivering no current and it is therefore operating as if on open circuit. The voltage at its terminals, v_s, is the open-circuit voltage. The voltage across the external resistor $v_R = iR_b = 0$ since $i = 0$. In this situation the external resistor could be disconnected from the rest of the network and the external source (Figure 3.5(c)) without affecting either of them. The voltage difference between the open terminals would

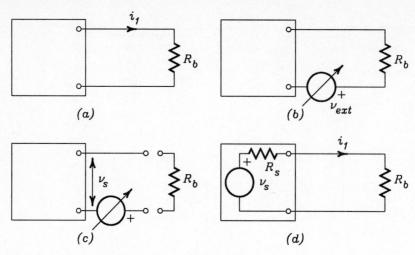

Figure 3.5

remain zero and an application of the principle of superposition will yield:

$$v_s - v_{ext} = 0 \quad \text{or} \quad v_s = v_{ext} \tag{3.1}$$

making clear that the external source has been set to the open-circuit voltage of the driving network. The total current is also amenable to an application of the principle of superposition when the effect of each part will be taken separately, the other being rendered ineffectual or dead. We have already said that the current delivered by the driving network with the external network dead is i_1. In the reverse situation the driving network is rendered ineffectual by suppressing all internal sources. The usual rules for combining resistors in series and parallel would then yield a single effectual value of resistance for that network, R_s, say. The total current in (b) may then be given as:

$$i = i_1 - \frac{v_{ext}}{R_s + R_b} = 0 \tag{3.2}$$

so that:

$$i_1 = \frac{v_{ext}}{R_s + R_b} = \frac{v_s}{R_s + R_b} \tag{3.3}$$

where we have made use of equation (3.1) to substitute for v_{ext}. Exactly the same result as regards i_1 would be obtained with R_b connected to a simple circuit (Figure 3.5(d)) comprising an ideal voltage source v_s and a series resistor R_s. That simple circuit is the Thevenin equivalent. The added adjustable external source was a stratagem for equating the characteristics of the actual network to its equivalent circuit.

If the external network itself contains sources, apart from the added source, then for the application of superposition we allow that it can deliver a current i_b when other parts of the network are dead. When the currents of the driving network and the added source are calculated, all sources in the external network must be

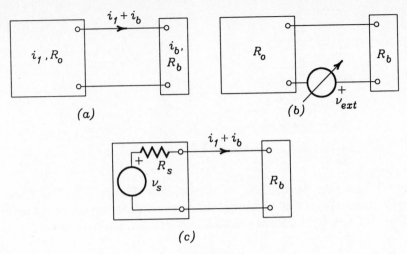

Figure 3.6

suppressed and so it will still be possible to characterize it with a single value of resistance R_b. We include i_b in equation (3.2) to obtain:

$$i = i_1 + i_b - \frac{v_{ext}}{R_s + R_b} = 0 \tag{3.4}$$

so that:

$$i_1 + i_b = \frac{v_{ext}}{R_s + R_b} \tag{3.5}$$

and

$$v_{ext} = i_1(R_s + R_b) + i_b(R_s + R_b) \tag{3.6}$$

The first term on the right-hand side of equation (3.6) is the open-circuit voltage of a source required to sustain a current i_1 in the interconnection between the two parts of the network when the sources in both the driving network and the external network have been suppressed. Using the symbol v_s for this voltage, as in the earlier proof for an external network containing no sources, and substituting in equation (3.5) we find:

$$i_1 + i_b = \frac{v_s}{R_s + R_b} + i_b$$

The current of the external network therefore appears as a term added to both sides of equation (3.3) which established Thevenin's theorem. By application of the principle of superposition we know that $i_1 + i_b$ is the total current which flows between the two parts of the original network (Figure 3.6(a)). We see that neither the total current nor the contribution to the total current of the external network (i_b) is changed when the external network is connected to the Thevenin equivalent of the driving network (Figure 3.6(c)). The availability of the current i_b does not affect the establishment of the Thevenin equivalent and the earlier restriction is removed.

Thevenin's theorem is central to the theme of network analysis and it is appropriate to illustrate it with a number of examples.

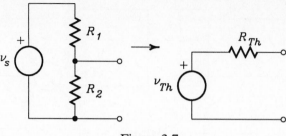

Figure 3.7

Example 3.3 First take the familiar potential divider (Figure 3.7). The Thevenin equivalent voltage (v_{Th}) is that which appears at the output terminals of the network when open-circuited. By the usual potential divider calculation we find:

$$v_{Th} = v_s \frac{R_2}{R_1 + R_2}$$

The Thevenin equivalent resistance is found with the source suppressed and is then the combination of two resistors in parallel:

$$R_{Th} = \frac{R_1 R_2}{R_1 + R_2}$$

Example 3.4 What might be called a cascaded potential divider (Figure 3.8) provides a useful illustration. The temptation is to suppress the source and attempt to find the effective resistance in one move, which is algebraically cumbersome. Instead we reduce the network piecewise, starting closest to the source (Figures 3.8(b) to (d)), reducing each potential divider in turn. The particular set of resistor values chosen for the example ensures that the Thevenin voltage is halved at each stage (a binary sequence). The arrangement is known as an R–2R ladder and is much used in circuits for converting digital signals to analogue form when taps are taken from all of the dividers.

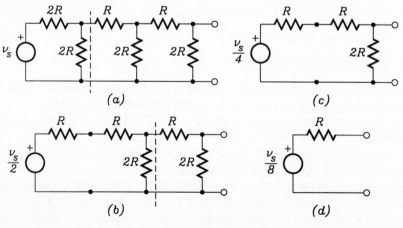

Figure 3.8

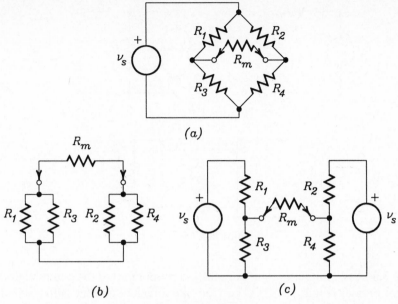

Figure 3.9

Example 3.5 The Wheatstone bridge (Figure 3.9(a)) is another familiar network which we might usefully analyze in this way. The off-balance current through the detector involves considerable calculation by any of our previous methods. Now we treat the detector as an external circuit and seek an equivalent circuit for the bridge driving it. With the source suppressed the equivalent resistance may be seen as a series–parallel combination (Figure 3.9(b)) so that:

$$R_{Th} = \frac{R_1 R_3}{R_1 + R_3} + \frac{R_2 R_4}{R_2 + R_4}$$

The equivalent voltage may be obtained by invoking one of the source translations described in Chapter 2 to produce the circuit of Figure 3.9(c) and then expressing the result as the superposition of two potential dividers:

$$v_{Th} = v_s \frac{R_3}{R_1 + R_3} + v_s \frac{R_4}{R_2 + R_4}$$

The off-balance current is therefore simply $v_{Th}/(R_{Th} + R_m)$ where R_m is the resistance of the detector.

A Norton equivalent circuit may be obtained from a Thevenin equivalent by a source transformation (Figure 3.10). It follows that if one can be proved then so can the other. An independent proof of Norton's theorem can be given in an analogous way to that for Thevenin's theorem: by introducing an adjustable current source into the external circuit but shunting it across the driving circuit (Figure 3.11). One adjusts the source for zero voltage across both networks, when it is as if they are short-circuited. The details of the proof are left to the reader.

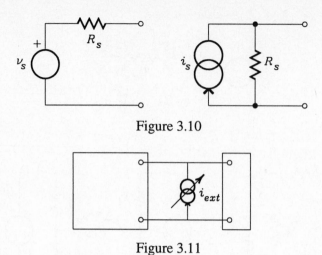

Figure 3.10

Figure 3.11

Example 3.6 As a simple example we take the counterpart of the potential divider, the current divider (Figure 3.12). The Norton equivalent current is that when the terminals are short-circuited:

$$i_N = i_s \frac{R_1}{R_1 + R_2}$$

The Norton equivalent resistance is calculated with the source replaced by an open circuit:

$$R_N = R_1 + R_2$$

Analogous expressions to those for the potential divider may be seen by working in terms of the conductance $G = 1/R$ rather than the resistance for the current divider. Then the Norton equivalent current:

$$i_N = i_s \frac{G_2}{G_1 + G_2}$$

and the Norton equivalent conductance:

$$G_N = \frac{G_1 G_2}{G_1 + G_2}$$

A current divider is perhaps more familiarly known as a meter shunt.

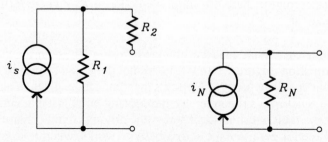

Figure 3.12

3.3   Power Transfer between Systems

Power calculations are discussed in detail in Chapter 9. Here we confine ourselves to some immediate observations on how the concept of an equivalent circuit can assist our understanding of how to couple networks together to obtain the best effect in terms of the power transferred. The particular networks may have been designed primarily for the transmission of power or for the transmission of information. In the former case the emphasis is on providing energy for other devices (as with a power supply). In the latter case we may speak of a signal power but the emphasis is on the transmission of information (as with a transducer).

For information handling, the best we can hope to do is to maximize the ratio of the signal power to the noise power (extraneous random effects), which will inevitably be added by the circuit. To maximize the ratio we must maximize the signal power transferred from one network to the next. For a network, represented by its Thevenin equivalent, driving a passive network represented by a simple resistor (Figure 3.13) the power transferred is:

$$P = i^2 R_L = v_s^2 \frac{R_L}{(R_s + R_L)^2}$$

As $R_L \to 0$, i increases but $P \to 0$ because the voltage across the load decreases. Further, as $R_L \to \infty$, the voltage across the load increases but again $P \to 0$ because $i \to 0$. Between these two limiting cases the power is a maximum given by the condition:

$$\frac{dP}{dR_L} = v_s^2 \left[\frac{1}{(R_s + R_L)^2} - \frac{2R_L}{(R_s + R_L)^3} \right] = 0$$

which occurs for:

$$\frac{2R_L}{R_s + R_L} = 1 \quad \text{that is} \quad R_s = R_L$$

This is the maximum power theorem.

The same criterion can easily be seen to be highly undesirable for power supplies. The power supply authority would clearly be dismayed if 100 W were to be developed in their equipment for every 100 W lamp lit in the supply area. In this case we require $R_L \gg R_s$ so that we work very well away from the maximum power

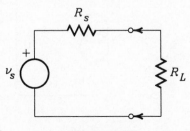

Figure 3.13

condition where equal power is developed in the source and the load. The condition is here one of maximum efficiency, a maximum of the power developed in the load compared with that developed in the source. The distinction between signals and power can become somewhat blurred as in the later stages of signal-handling equipment. For instance, in the power-output stages of a hi-fi amplifier there would be a requirement for maximum efficiency, not maximum power transfer.

3.4   Network Duals

We are all the time seeking ways of reducing the number and complexity of circuits with which we have to deal. Equivalent circuits have been introduced to reduce complexity and we find that source transformation halves the total number of these that need be maintained. Still confining our attention to purely resistive networks, the possibility of further economy may be seen by writing generalized mesh-current (KVL) and node-voltage (KCL) equations, both simple sums, as follows:

$$\sum i_m R_b + \sum v_s = 0$$
$$\sum v_r G_b + \sum i_s = 0$$

where i_m and v_r are the mesh currents and node voltages respectively, and R_b and G_b are the branch resistances and conductances respectively. The source terms which cannot be re-expressed in terms of mesh currents and node voltages have been summed separately. If we now envisage a bilateral (two-way) translation between voltage and current and between resistance and conductance, then the mesh equations of one network will become the node equations of another, while node equations will become mesh equations. Networks with their node and mesh equations related in this way are said to be *duals*. A solution of the mesh equations of one is therefore a solution of the node equations of the other.

Example 3.7 The potential divider (Figure 3.14(a)) may once again be used to provide a simple illustration. For the potential divider the mesh equation and its dual are:

$$v_s - iR_1 - iR_2 = 0 \qquad i_s - vG_1 - vG_2 = 0$$

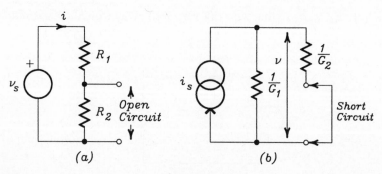

(a)   (b)

Figure 3.14

The second equation is the node-voltage equation of the circuit in Figure 3.14(b), the current divider. We therefore see another connection between the potential divider and the current divider: they are duals. The node-voltage equation of the potential divider is almost trivial, but for completeness it and its dual equation are:

$$\frac{v_{R_1}}{R_1} + \frac{v_{R_2}}{R_2} = 0 \qquad \frac{i_{G_1}}{G_1} + \frac{i_{G_2}}{G_2} = 0$$

The second is a mesh-current equation for the current divider.

Bear in mind that voltage and current are numerically the same in the potential divider and its dual, as are resistance (R) and conductance (G). Therefore for $R_1 = R_2 = 10\ \Omega$ (say), $G_1 = G_2 = 10$ S and the resistances in the current divider are both $0.1\ \Omega$. A 12 V source applied to the potential divider will produce a current of 0.6 A while the 12 A source in the dual will develop 0.6 V across the two resistors in parallel.

The duality of other quantities has also emerged. A short circuit needs to be translated to an open circuit, a series connection to a parallel connection and a node to a mesh. From what we already know of the behaviour of capacitors and inductors we can see that they are duals also:

$$v_L = L\frac{di}{dt} \qquad i_C = C\frac{dv}{dt}$$

After translating voltage to current and vice versa the equations become the same if inductance is translated to capacitance and vice versa. By exchanging current for voltage, short circuit for open circuit, and so on, Norton's theorem is produced from Thevenin's theorem and so we see that the theorems themselves are duals. However, it must be made clear that while the theorems are duals, the circuits produced by the application of the theorems to a particular network are certainly not duals. Our simple model of a car battery, introduced in Chapter 1, is used to illustrate this point in Figure 3.15 where the Thevenin and Norton equivalents of the battery and their duals are shown. A mesh equation may be written for each Thevenin equivalent by introducing a short circuit at the output terminals. The duals of the mesh equations are the node equations for the Norton equivalents where the open circuits at the output terminals are the duals of the short circuits introduced for the Thevenin circuits.

In more complicated situations, graph theory, introduced in the last chapter, is indispensable in obtaining the dual network. One places a point inside every mesh of the graph and a further point in the 'external mesh', that is, outside of the whole network. The points are joined so that only one line crosses every branch of the original graph. For the simple case of the potential divider (Figure 3.16) it will be seen that the dual graph is that for the current divider. The two circuits have already been recognized as duals. Further, as a non-planar graph has no unique representation, so it and its associated circuit have no unique dual. Duality is therefore confined to planar circuits.

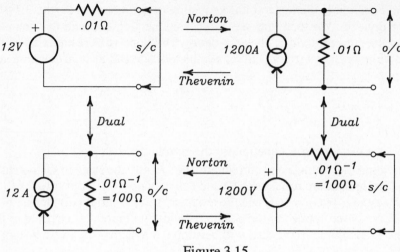

Figure 3.15

Summary

The principle of superposition, which formally underpins the linearity of systems, has been re-examined, on this occasion in regard to its usefulness in analyzing networks where more than one source is present. A partial solution for a network, the voltage across a single branch or the current in such a branch, can be found by adding the effects of each source with all the other sources suppressed. Single branches of particular interest are those containing the input or the output with the remaining network adequately represented by a single resistor and possibly a single source connected internally between the input or the output terminals. The representation of a network with respect to a chosen pair of terminals, the input or the output, in such a minimal way is its so-called *equivalent circuit*. *Thevenin's theorem* and *Norton's theorem* provide a formal means for reducing a network to its equivalent circuit, the former to a voltage source and series resistor, the latter to a current source and shunt resistor.

Provided with the idea of the equivalent circuit it is a simple matter to decide how best to transfer the maximum power to (or from) an external circuit. The result

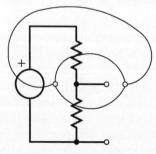

Figure 3.16

is the *maximum power theorem* where we see that the source and load resistors should be the same for the maximum transfer of power.

Lastly we noted that if Kirchhoff's voltage law and Kirchhoff's current law equations are written in a general way with the sources summed separately then an algebraic manipulation may be envisaged which translates voltage to current and resistance to conductance, and vice versa. The Kirchhoff voltage law equations are translated to Kirchhoff current law equations, and vice versa. The networks represented by the new sets of equations produced in this way are said to be the *duals* of the originals. Duality is in essence another way of reducing the number of different networks that may be enumerated. Thevenin's and Norton's *theorems* could be seen as duals one of the other (but Thevenin and Norton reductions of the same network are *not* duals one of the other).

Problems

3.1 Find, by the application of the principle of superposition and appropriate network transformations, the voltage v_0 in the circuit shown in Figure 3.17.

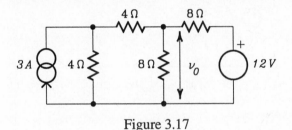

Figure 3.17

3.2 Find, by the application of the principle of superposition, the voltage v_0 in the circuit shown in Figure 3.18 ($2v_1$ is a controlled source). This network has been studied previously using other methods (Problem 2.5) and contributions to the present solution should also already have been obtained (Problem 2.3). Compare the solutions obtained by the various methods.

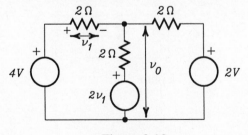

Figure 3.18

3.3 Derive *both* Thevenin and Norton equivalents for the circuit shown in Figure 3.19.

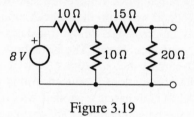

Figure 3.19

3.4 Derive *both* Thevenin and Norton equivalents for the circuit shown in Figure 3.20.

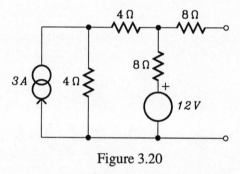

Figure 3.20

3.5 Draw the network which is dual to that shown in Figure 3.21 and evaluate the component values. Confirm that the Norton and Thevenin equivalents of the given network and its dual respectively are themselves dual.

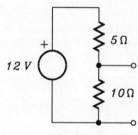

Figure 3.21

4

Networks with Inductors and Capacitors

4.1 General n-terminal and Two-port Networks

Before including capacitors and inductors in our consideration of networks and so opening up the major part of the topic of network analysis, it is appropriate to introduce the *two-port network* as a basis for our future discussion. The two-port, as its name implies, has one pair of input terminals and one pair of output terminals (Figure 4.1). It acknowledges the situation, so often encountered in practice, of a network producing a single output at one pair of terminals for an input applied at the only other pair of terminals. It may contain dependent (or controlled) sources but it contains no independent sources, except those implicit in the presence of the controlled sources (an active element includes its power supply; Chapter 1). All other independent sources are external to the network. As a consequence, the current at each terminal of a terminal pair (port) is equal but opposite to that at the other terminal of the pair. Some common examples of two-ports are shown in Figure 4.2. If the restriction on the inclusion of independent sources is removed then the current at the four terminals will, in general, be different and we are dealing with a general four-terminal network which itself is a special case of the general n-terminal network.

The use of two-ports has been implicit in much of our previous reasoning and is exemplified by the potential divider (where one input terminal and one output terminal are connected). A common (but incomplete) specification of the properties of a two-port is given in its *transfer function*, so-called because it describes how the excitation at the input is transferred to the output as a response. A voltage or a current may be taken as the excitation and a voltage or a current identified as

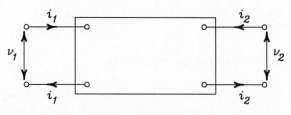

Figure 4.1

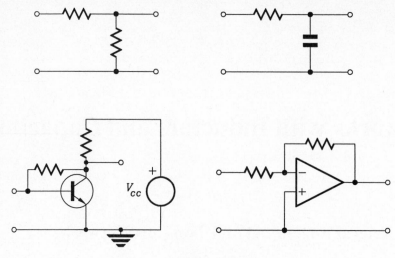

Figure 4.2

the output. It follows that there are four possibilities: a voltage transfer function and a current transfer function (both dimensionless), a voltage-to-current ratio and a current-to-voltage ratio. When extended to other than purely resistive circuits the last two have the dimensions of *impedance* and *admittance* respectively, of which more in section 8.3. The transfer functions are not independent of the source and load characteristics, which is where they are incomplete in their specification. Implicit in all that we do will be the application of an ideal source to the input with the output left on open circuit. A more complete description relates the currents and voltages at both the input and the output via a set of two-by-two matrices, the *two-port parameters*, but that goes beyond the scope of this book.

Another situation of practical significance is the use of one or a combination of components to determine that a particular current will be drawn from a voltage source or that a particular voltage will be developed across a current source. In this case there is not an input and an output but just one pair of terminals and we have a *one-port*. If we identify the voltage as the excitation then the current is the response, the ratio is an impedance. The reverse situation produces a ratio which is an admittance. Both ratios are said to be *driving-point functions*. However, it might be remarked that to verify the current–voltage relationship in either case it would be necessary to introduce a second pair of terminals. In the standards laboratory one finds this as a permanent feature of components so that even a simple resistor is provided with four terminals (the four-terminal resistor), one pair for the current and the other pair for the voltage measurement.

4.2 The Free Response of Systems

For the purely resistive networks to which we have so far confined ourselves, the node-voltage (KCL) and mesh-current (KVL) methods produce sets of coupled linear algebraic network equations. The output of such systems will mimic the time behaviour of the input and in addition, if the system is passive, the output can never exceed the input. When energy storage elements, capacitors and inductors are included, the above methods will produce linear differential equations as indicated in Chapter 2. The network equations can be combined to produce a so-called system differential equation. The charge transferred between the plates of a capacitor will determine the potential difference across the capacitor. The current is therefore related to the time-rate-of-change of the potential difference. In a similar way, the rate-of-change of flux in an inductor is related to the potential difference across the inductor (Figure 1.4). As a consequence the output of a system containing such elements will no longer follow the input as a function of time.

Systems where energy storage elements have the time-shifting effect are very much part of our common experience where mechanical systems are concerned. A motor-car suspension which has no springs is known to be very uncomfortable as there is total and immediate response to every bump in the road, but it would give a very even ride on a totally smooth road. The addition of springs means that the bumps are evened-out but the response continues for some time. A too-soft suspension may produce its own discomfort. Direct analogies can be seen between the spring constant of a mechanical system and the capacitance of an electrical system, also between mass and inductance. This once again makes the point that electrical networks are physical systems and that we are not generating fundamentally new laws or methods to deal with them.

At the risk of overworking the analogy we once again refer to a motor-car suspension. It is common practice to press down on the car bumper then release it and watch how the car bounces with a view to testing the suspension and so decide how it will perform on the road. A rudimentary test of many mechanical systems can be effected by administering a judicious thump. There are evidently certain inherent properties of any mechanical system which are central to how it will perform under stimulation. The same is true for electrical systems and it is for this reason that we start with a consideration of the free, or natural, response.

Consider first a voltage source connected to a resistor and capacitor in series (an RC network), as shown in Figure 4.3. An application of Kirchhoff's voltage law yields the following:

$$R\frac{dq}{dt} + \frac{q}{C} = x(t) \tag{4.1}$$

$$\frac{q}{C} = y(t) \tag{4.2}$$

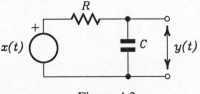

Figure 4.3

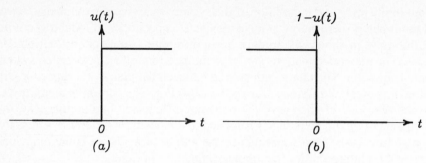

Figure 4.4

Substituting both equation (4.2) and its differential into equation (4.1) produces:

$$CR\frac{dy(t)}{dt} + y(t) = x(t) \tag{4.3}$$

the system differential equation. A full description is that it is a linear first-order non-homogeneous differential equation. It is linear because there are no terms of higher than the first power, first-order because that is the maximum order of the differential terms and non-homogeneous because there are terms in both x and y. The highest-order term in $y(t)$ is taken to characterize the system and so we recognize the RC circuit as a *first-order system*. In general, the right-hand-side of the equation, known as the *forcing function*, will be a linear combination of $x(t)$ and all its differential coefficients. We shall not attempt a complete solution of the equation until section 4.4.

At this point, as we appear to be becoming more deeply immersed in differential equations, it might be remarked that our ultimate objective is to avoid not only having to solve differential equations but even having to write them down in the first place. However, the saving of effort later will not be appreciated unless what otherwise needs to be done is first examined. For the moment we require a solution in the special case where the input has been removed so the forcing function is zero and the system is exhibiting its free behaviour. Of course, if the input has always been zero then the system will be completely mute and of no interest. The situation we envisage is one in which the input having previously been non-zero is removed, at $t = 0$, say. The system is therefore left with initial stored energy at $t = 0$. We may describe this initialization of the system in a precise way using one of the so-called singularity functions which find so much use in the subject. The unit step function (Figure 4.4(a)) $u(t)$ is zero for $t < 0$ and unity for $t > 0$. The function $v_0(1 - u(t))$ (Figure 4.4(b)) describes the behaviour we require where v_0 is the voltage maintained across the capacitor prior to the time $t = 0$.

The system equation with $x(t) = 0$ is now a homogeneous equation:

$$CR\frac{dy}{dt} + y = 0 \tag{4.4}$$

which is simple enough to integrate directly after first rearranging to:

$$\frac{dy}{dt} = \frac{-1}{CR}y \tag{4.5}$$

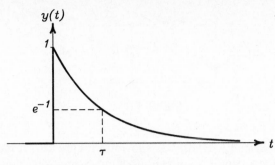

Figure 4.5

or
$$\frac{dy}{y} = \frac{-1}{CR}dt$$

Hence
$$\int \frac{dy}{y} = \frac{-1}{CR}\int dt$$

and
$$\ln y = \frac{-1}{CR}t + K$$

The constant of integration, K, is determined by the fact that $y = v_0$, say, for $t < 0$ and remains at this value for an infinitesimal interval at $t = 0$, as the charge on a capacitor cannot change instantaneously; consequently $K = \ln v_0$.

So
$$\ln y - \ln v_0 = \frac{-1}{CR}t$$

$$\ln\left\{\frac{y}{v_0}\right\} = \frac{-1}{CR}t$$

Hence
$$y = v_0 e^{\frac{-1}{CR}t} \qquad (4.6)$$

which is the familiar exponential decay shown in Figure 4.5. The time at which the voltage falls to $1/e$ of its initial value is regarded as characteristic of the decay and is seen to be CR, called the *time constant*, τ, of the decay.

Example 4.1 The combination of, for instance, a 1 μF $(= 10^{-6}$ F) capacitor and a 1 MΩ $(= 10^6 \ \Omega)$ resistor can readily be seen to produce a 1 second time constant:

$$\tau = CR = 10^{-6}10^6 = 1$$

at which time the original voltage to which the capacitor was charged will have dropped to $1/e = 63\%$. The value of 63% can be used to recognize the time corresponding to the time constant and is 'one of those things people remember'. The values chosen above are probably at the high end of values commonly employed in circuit construction, discounting electrolytic capacitors, but are always readily available.

We now turn to the simplest circuit containing the two different types of energy storage element by considering an inductor and a capacitor connected in series to a

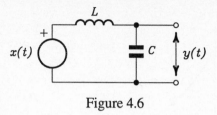

Figure 4.6

voltage source (the series LC circuit) as shown in Figure 4.6. It must be admitted at the outset that such a circuit is something of an abstraction as the only available way of completely eliminating resistance – superconductivity – cannot be put to everyday use. However, the properties of this circuit set a limit on the behaviour of practical circuits. Again the application of Kirchhoff's voltage law yields:

$$L\frac{di}{dt} + \frac{q}{C} = x(t)$$

or $\qquad L\frac{d^2q}{dt^2} + \frac{q}{C} = x(t)$ $\qquad\qquad\qquad\qquad\qquad$ (4.7)

while $\qquad\qquad y(t) = \frac{q}{C}$ $\qquad\qquad\qquad\qquad\qquad\qquad$ (4.8)

Substituting both equation (4.8) and its second differential into equation (4.7) produces:

$$LC\frac{d^2y(t)}{dt^2} + y(t) = x(t) \qquad\qquad\qquad (4.9)$$

The system differential equation in this case is a linear second-order non-homogeneous equation. The LC circuit is accordingly described as a *second-order system*. The free response may again be examined by setting the input to zero having first provided the system with some initial stored energy by the application of the function $v_o(1 - u(t))$. That is, the corresponding homogeneous equation is:

$$LC\frac{d^2y(t)}{dt^2} + y(t) = 0 \qquad\qquad\qquad (4.10)$$

Rearranging equation (4.10) we obtain the well-known equation of simple harmonic motion:

$$\frac{d^2y}{dt^2} = \frac{-1}{LC}y$$
$$= -\omega_0^2 \qquad\qquad\qquad\qquad (4.11)$$

where $\omega_0^2 = 1/LC$. A specific solution in this case is provided by the sine (or cosine) function:

$$y = v_0 \cos\omega t \qquad\qquad\qquad\qquad (4.12)$$

Example 4.2 It is clear from the expression $\omega_0^2 = 1/LC$ that for high values of ω_0 both L and C will be small. Small-value inductors and capacitors therefore find application in circuits operating at radio frequencies, except that here digital techniques for frequency synthesis and demodulation are now so often employed. Elsewhere in circuit construction is a general tendency to avoid the use of inductors, because they are bulky components, and to use instead active circuits employing capacitors which behave as inductors. However, a common situation where modest-sized inductors are to be found is in mains filters used to prevent signals entering or exiting a piece of equipment along the mains lead. Typical values for the inductor used would be 10 mH and for the capacitor 1000 pF. The two components would 'ring' (have a free response) at a frequency (f_0) given by:

$$f_o = \frac{1}{2\pi\sqrt{1000\ 10^{-12}10\ 10^{-3}}} \simeq 50\ \text{kHz}$$

As we have already mentioned, a system which is characterized by a particular free response will respond readily when that oscillation is applied as a stimulus. As a consequence, the above component combination will have the effect of 'shorting-out' inputs at 50 kHz and a wide range of frequencies around that frequency. It is the basis of the design of a mains (or line) filter where attenuation of 'common mode' interference is required; other aspects of the design ensure that desired frequencies (the differential mode) are not attenuated.

4.3   The Role of the Sine Function

We have now made our first contact with the sine function in regard to describing time-varying voltages and currents in networks. (In this context we use sine as a generic term to include the cosine function.) The simple solution just given might be seen by the reader as an early justification for the known preoccupation in this field with trigonometric quantities to describe variation with time. Such a conclusion would be premature. Certainly such a preoccupation with sinusoidally (or cosinusoidally) varying voltages and currents is easy to understand in the case of electrical engineering where the emphasis in the use of networks is for the transmission of power. In electronic engineering, on the other hand, the emphasis in the use of networks is on signal handling, and signal sources never emit pure sinusoids. A pure, uninterrupted sinusoid carries no information. Often an instrument described as a signal generator may in fact be no more than a sinusoidal waveform generator. To illustrate the point that even a fairly primitive signal is far from being sinusoidal, the representation of a simple message, the greeting 'hello', is shown in Figure 4.7. It is a photograph of an oscilloscope trace of the output of a microphone and amplifier. The general nature of such everyday signals as music or television transmissions will be well known to the reader and serves to underline the point. Nevertheless, the same preoccupation with sinusoids as obtains for power-handling circuits is usually evident in the treatment of networks when they are intended for information handling. Complete justification for this approach will be given in due course, suffice it to say now that a most important mathematical theorem given by Fourier in 1807 specifies how any function may be expressed as a sum of harmonic sine and cosine functions. A convincing demonstration of the application of this theorem can be given using an instrument called a Fourier waveform analyzer. The result of superimposing odd harmonics with specified amplitudes up to the ninth is

Figure 4.7

shown in Figure 4.8. A reasonable facsimile of a square wave is obtained. Taking very many more harmonics will produce a much better square wave.

Fourier's theorem becomes applicable to linear circuits by virtue of the fact that sine waves are the only waveforms transmitted by a linear circuit without change. That is to say, a sine wave may be changed in amplitude and displaced in time (phase shifted) as shown in Figure 4.9 but it remains functionally a sine wave.

The basis for the simplest representation of sinusoidally varying quantities (voltages, currents, displacements and so on), and all schemes which derive from it, is the idea of a line, one end of which is fixed at the origin of axes while the other end describes a circle, the line sweeping-out an angle of ω radians per second (Figure 4.10). The projection of the line on the x-axis is $v_x = v_o \cos \omega t$, where v_o is the length of the line and the amplitude of the oscillation. In this case the choice of symbol indicates a voltage. The line has a magnitude and direction associated with it and so may correctly be described as a vector. The representation we have invoked is therefore that of the rotating vector. At this point it is worth remarking that it is only the *representation* that is vectorial; we are saying nothing about the directional properties of the quantity represented. If instead we choose to project

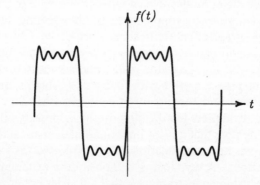

Figure 4.8

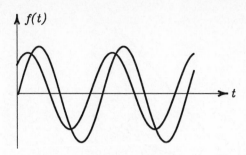

Figure 4.9

onto the y-axis, we obtain $v_y = v_o \sin \omega t$ which is the same as $v_y = v_o \cos(\omega t - \pi/2)$. That is, the projection on the y-axis is the same as the projection on the x-axis for a vector starting its rotation from $-90°$. We say that two oscillations, represented by the projections on the x-axis of two vectors making an angle of 90°, are *in quadrature*, or more commonly, that they are 90° *out of phase*. Phase, or phase angle, is the general term we shall use from now on to describe the angle swept out by the rotating vector. In fact what are of greatest interest are *differences in phase angle* between various oscillations and how these are changed by particular circuits. There are various notable out-of-phase conditions, the one we have already met, 90° $(\pi/2)$ or $\frac{1}{4}$ cycle and 180° (π) or $\frac{1}{2}$ cycle. To combine sinusoidal oscillations, given that each may be represented by a vector, the rules of vector algebra may be applied to vectors in a two-dimensional surface.

Example 4.3 Let us find the resultant of $v_1 = v_{01} \cos \omega t$ and $v_2 = v_{02} \cos(\omega t + \Phi)$. For any particular instant in time the two may be combined in a vector triangle as shown in Figure 4.11. The resultant (v_r) is the projection of the hypotenuse of the triangle on the x-axis and may be written as $v_0 \cos(\omega t + \delta)$. Expressions for v_0 and δ in terms of v_{01}, v_{02} and Φ may be obtained by first dropping a perpendicular onto the extension of v_{01}, as shown. Simple geometry yields the result that:

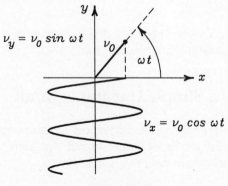

Figure 4.10

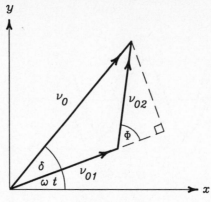

Figure 4.11

$$v_0 = \sqrt{(v_{01} + v_{02}\cos\Phi)^2 + (v_{02}\sin\Phi)^2} \qquad (4.13a)$$

$$\text{and} \quad \tan\delta = \frac{v_{02}\sin\Phi}{v_{01} + v_{02}\cos\Phi} \qquad (4.13b)$$

Alternatively, a purely algebraic method may be adopted in which various trigonometric identities are employed to obtain the result as follows:

$$
\begin{aligned}
v_r(t) &= v_{01}\cos\omega t + v_{02}\cos(\omega t + \Phi)\\
&= v_{01}\cos\omega t + v_{02}(\cos\omega t\cos\Phi - \sin\omega t\sin\Phi)\\
&= (v_{01} + v_{02}\cos\Phi)\cos\omega t - v_{01}\sin\Phi\sin\omega t
\end{aligned}
$$

The square root of the sum of the squares of the coefficients of $\cos\omega t$ and $\sin\omega t$ is v_0 as above. Multiplying and dividing the last result by v_0 gives:

$$
\begin{aligned}
v_r(t) &= v_0\left\{\frac{(v_{01} + v_{02}\cos\Phi)}{v_0}\cos\omega t - \frac{v_{02}\sin\Phi}{v_0}\sin\omega t\right\}\\
&= v_o\{\cos\delta\cos\omega t - \sin\delta\sin\omega t\}\\
&= v_0\cos(\omega t + \delta)
\end{aligned}
$$

with $\tan\delta$ again given as above. The method may be extended to the combination of any number of vectors with proportionally more work.

4.4   Response to a Simple Harmonic Input

The solution of the non-homogeneous differential equation describing the behaviour of a circuit with non-zero input is the sum of two parts, *the complementary function*, $y_c(t)$ and *the particular solution*, $y_p(t)$, that is:

$$y(t) = y_c(t) + y_p(t) \qquad (4.14)$$

The first part is the solution of the homogeneous equation, which we have already encountered in regard to the free response. Its role here is to account for the transition which must occur in the state of the output as we change the initial conditions by applying the input, often called the *switching transient*. (In the case of the free response the initial conditions were changed by *removing* the source of excitation.) The second part represents the steady response of the circuit after the switching transient has passed. It is so named because it describes the response to the *particular* input to the circuit.

For linear differential equations with constant coefficients, such as concern us here, a completely general solution is, in principle, always possible. The method is based upon the direct integration of first-order equations which produces both the particular integral and the complementary function. Integration of the first-order equation:

$$\frac{dy}{dt} + ay = x$$

is made possible by first multiplying throughout by e^{at}:

$$e^{at}\frac{dy}{dt} + ae^{at}y = e^{at}x$$

which may be rewritten:

$$\frac{d}{dt}(e^{at}y) = e^{at}x \tag{4.15}$$

Integrating we obtain:

$$e^{at}y = \int e^{at}x\,dt + c$$

or $\qquad y = ce^{-at} + e^{-at}\int e^{at}x\,dt \tag{4.16}$

One attempts to solve equations of higher order by using the first-order solution as a model and employing successive integration or by factorizing into first-order equations which are then integrated. It would be out of place to continue such a general treatment here. The basis of the method, as applied to the RC circuit, is given in Appendix B. Various less general methods having different realms of suitability are available. One such which is applicable to a large class of forcing functions, including all those of interest to us here, is the *method of undetermined coefficients*. The particular integral is taken as a linear combination of the forcing function and all its derivatives. The complementary function is obtained separately from a solution of the homogeneous equation.

Example 4.4 The method may be adequately illustrated by reference to the RC circuit of Figure 4.3 with the input $\cos \omega t$. It is not an oversimplification to consider such an input, since from our brief contact with Fourier's theorem we know that combinations of such simple functions will produce any desired input. The equation we require to solve is therefore:

$$CR\frac{dy(t)}{dt} + y(t) = \cos \omega t \tag{4.17}$$

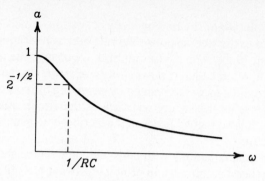

Figure 4.12

We take:

$$y_p(t) = A\cos\omega t + B\sin\omega t \tag{4.18}$$

It is a simple trigonometric manipulation to show that equation (4.18) is the same as:

$$y_p(t) = a\cos(\omega t - \delta)$$

where $a = \sqrt{(A^2 + B^2)}$ and $\tan\delta = B/A$. The supposed solution therefore allows that the output would be changed in amplitude and phase but not in frequency, as common sense would dictate. Indeed, such an approach (guessing) is a perfectly acceptable means of finding a solution, the only requirement is that the result *is* a solution. Substituting equation (4.18) into equation (4.17):

$$CR(-A\omega\sin\omega t + B\omega\cos\omega t) + A\cos\omega t + B\sin\omega t = \cos\omega t$$

Equating coefficients of $\sin\omega t$ and $\cos\omega t$:

$$B\omega CR + A = 1 \tag{4.19a}$$
$$B - A\omega CR = 0 \tag{4.19b}$$

Squaring and adding we obtain:

$$(A^2 + B^2)(1 + \omega^2 C^2 R^2) = 1$$

So that $a = \sqrt{A^2 + B^2} = \dfrac{1}{\sqrt{1 + \omega^2 C^2 R^2}}$ $\qquad$ (4.20a)

and $\qquad \tan\delta = \dfrac{B}{A} = \omega CR$ $\qquad$ (4.20b)

The variation of a with frequency is an important property of the circuit which we will encounter again later on. For the moment it will be useful to have a graphical record of the result and this is given in Figure 4.12.

The complementary function (solution of the homogeneous equation) in this case is:

$$y_c(t) = K_1 e^{\frac{-1}{CR}t}$$

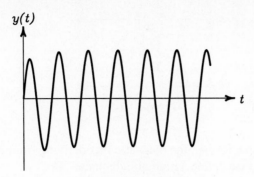

Figure 4.13

where K_1 is a constant of integration related to that introduced in obtaining the free response by:

$$\ln K_1 = K$$

The general solution for the input $u(t)\cos\omega t$, that is a cosine function starting at $t = 0$, is therefore:

$$y(t) = K_1 e^{\frac{-1}{CR}t} + \frac{1}{\sqrt{1+\omega^2 C^2 R^2}}\cos(\omega t - \delta)$$

The constant K_1 is determined *for a system with no initial stored energy* by again calling on the requirement that the charge on a capacitor may not change instantaneously, therefore $y = 0$ for an infinitesimal interval at $t = 0$. It follows that:

$$K_1 = \frac{-1}{\sqrt{1+\omega^2 C^2 R^2}}\cos(-\delta)$$

$$\text{and} \quad y(t) = \frac{1}{\sqrt{1+\omega^2 C^2 R^2}}\left[\cos(\omega t - \delta) - e^{\frac{-1}{CR}t}\cos(-\delta)\right]$$

The response is shown in Figure 4.13 where the transient, before the steady output is established, may be clearly seen.

It may be remarked that the above is the effort required to obtain a solution by classical methods for one of the simplest of all circuits. The need for a more streamlined approach is surely now evident to the reader.

4.5   Mutual Inductance: Networks with Transformers

The inductance (L) we have introduced thus far relates the self-linkage of flux by a coil to the current which sustains the flux:

$$NAB = Li \tag{4.21}$$

where N is the number of turns, A is the cross-sectional area and B is the magnetic flux density. When two coils are mounted in close proximity, usually by having

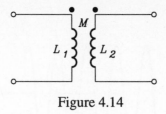

Figure 4.14

been wound on the same former, the magnetic flux deriving from the current in one will link the other and we define a mutual inductance (M):

$$N_2 A B_1 = M i_1 \tag{4.22}$$

where we have the same parameters as used previously but now identified by sub-scripts for the two coils. To distinguish the case of a single coil we should more properly call L the self-inductance. We take it for granted that M is symmetrical in the way that it connects the two coils, that is, we can also write:

$$N_1 A B_2 = M i_2$$

Such multiple winding components are commonly described as transformers, particularly when intended for use at low and audio frequencies, otherwise they are simply 'coils'. They are almost always manufactured using a machine which lays down multilayered windings all in the same sense, both 'up' and 'down' the coil-former, often arranging that the terminations to the windings all emerge at one end as a sheaf of wires. The symbolic representation of the component (Figure 4.14) gives no indication, without further annotation, as to its physical construction. It is ambiguous in regard to the sense of the windings and hence as to the expected polarity of the induced voltages. A convention for relating the connections of the actual device and its symbol is therefore adopted. One terminal of each coil in the symbol is identified by a dot. The dot indicates that *when current is into that terminal, and increasing, the voltage at the other dotted terminal has the same polarity as that arising from the self-inductance of the other winding,* that is, positive. It may be a matter of close examination, or even experimentation, to decide what should be identified as the dotted terminals of the device.

In almost all devices identified as transformers the windings will intentionally have been placed in close proximity so as to ensure that all the magnetic flux of one winding links the other, and vice versa. For a magnetic flux sustained by a current in winding 1 we would therefore have the ratio of the flux linkages:

$$\frac{N_1 A B_1}{N_2 A B_1} = \frac{L_1 i_1}{M i_1}$$

From which it follows that:

$$\frac{N_1}{N_2} = \frac{L_1}{M} \tag{4.23}$$

Equally for a flux sustained by a current in winding 2 we would find:

$$\frac{N_2}{N_1} = \frac{L_2}{M} \tag{4.24}$$

We multiply corresponding sides of equations (4.23) and (4.24) to produce, by further simple rearrangement:

$$M^2 = L_1 L_2$$

Dividing equation (4.24) by equation (4.23) we obtain, for the case of perfect coupling:

$$\frac{L_2}{L_1} = \left(\frac{N_2}{N_1}\right)^2 = n^2$$

where n is the turns ratio. When there is incomplete flux linkage between coils $Mi < Li$ with the result that:

$$\frac{N_1}{N_2} < \frac{L_1}{M} \quad \text{and} \quad \frac{N_2}{N_1} < \frac{L_2}{M}$$

and consequently:

$$M^2 < L_1 L_2$$

To return to an identity we write:

$$M = k\sqrt{L_1 L_2}$$

by introducing the coefficient of coupling, k. Numerically $0 \le k \le 1$.

The practical importance of coupled coils is seen by turning from their properties in terms of the static flux linkage to the situation when the flux linkage changes with time. Lenz's law relates the voltage induced in a coil to the changing magnetic flux linkage of the coil. Hitherto this has been expressed for the self-inductance, by differentiating equation (4.21):

$$v(t) = \frac{d(NAB)}{dt} = L\frac{di}{dt}$$

When the flux linked is sustained by a coupled coil there will be an induced voltage given in terms of the mutual inductance, as seen by differentiating equation (4.22):

$$v_m(t) = \frac{d(N_2 A B_1)}{dt} = M\frac{di_1}{dt}$$

For a pair of coupled coils each energized by an ideal voltage source (Figure 4.15) an application of KVL to each circuit produces the following set of system differential equations:

$$v_1(t) = L_1\frac{di_1}{dt} + M\frac{di_2}{dt} \tag{4.25a}$$

$$v_2(t) = M\frac{di_1}{dt} + L_2\frac{di_2}{dt} \tag{4.25b}$$

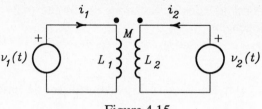

Figure 4.15

Example 4.5 A situation of practical significance is where a source is connected to just one winding (the primary) and we wish to know the no-load voltage available across the other winding (the secondary). By adjusting a source v_2 in the secondary circuit so that it is equal in magnitude but opposite in phase to the voltage produced by the secondary, the current i_2 will be zero and the magnitude of v_2 will be that of the open-circuit (no-load) voltage. In these circumstances equations (4.25a) and (4.25b) become:

$$v_1(t) = L_1 \frac{di_1}{dt}$$
$$v_2(t) = M \frac{di_1}{dt}$$

So that:

$$\frac{v_2(t)}{v_1(t)} = \frac{M}{L_1}$$

and for the case of a unity-coupled ($k = 1$) transformer we have, using equation (4.23):

$$\frac{v_2(t)}{v_1(t)} = \frac{N_2}{N_1} = n$$

The ratio of the magnitudes of the voltages at an instant in time is therefore simply the turns ratio. Depending upon whether n is greater or smaller than 1 we have, respectively, either a step-up or a step-down transformer.

Summary

As the basis for future discussion we introduced the idea of a two-port network, which is a special case of a general n-port network. The two-port, as its name implies, has just two pairs of terminals and is subject to the restriction that it contains no independent source (except those required to energize controlled sources). The arrangement accords closely with the practical situation of networks having one pair of terminals which accept an *input*, an *output* appearing at the other pair of terminals. An important connection between the excitation at the input and the response at the output is the *transfer function*.

Sometimes the important property of a network is that it determines a particular current for a particular applied voltage. Only one pair of terminals is required and we have a *one-port* for which we define *driving-point functions*, either impedance or admittance.

Now, introducing capacitors and inductors into simple circuit configurations with resistors we obtained, by an application of Kirchhoff's voltage law and Kirchhoff's current law methods, simple differential equations governing the behaviour of the networks. The equations are *linear*, reflecting the linearity of the systems with which we deal. The *order* of the equations is related to the number of energy-storage components, L and C, present. Imagining the circuits to have been energized for some time, we removed the stimulus and solved the equations to obtain the *free response* of a system. An LC circuit was seen to have a *sinusoidal* free response which prompted a review of the rotating-vector representation of simple harmonic motion together with the trigonometric methods for combining simple harmonic functions.

A solution of the differential equation for a simple network was performed to obtain the forced response to a simple harmonic input. It is apparent that the effort required to obtain a solution in the simplest situations suggests we seek more streamlined methods. In fact such methods will not only bypass the need to solve network differential equations but also render unnecessary our having to write them down in the first place.

The possibility of coupling between adjacent inductors was described by the introduction of *mutual inductance* in components known as *transformers*. For unity-coupled transformers the secondary to primary voltage ratio was seen to equal the turns-ratio.

Problems

4.1 Obtain the system differential equation for the network shown in Figure 4.16. Find the response of the system to the application of the input $x(t) = v_0(1 - u(t))$, that is, find the free response.

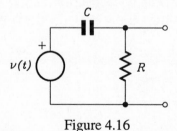

Figure 4.16

4.2 Solve the system differential equation for the circuit in Figure 4.16 to find the response to the input $x(t) = \sin \omega t$.

4.3 Obtain the system differential equation for the network shown in Figure 4.17. Show that the free response is given by $y = e^{-\sigma t}\cos\omega_m t$ where

$$\sigma = \frac{R}{2L} \quad \text{and} \quad \omega_m = \sqrt{\frac{1}{LC} - \frac{R^2}{4L^2}}$$

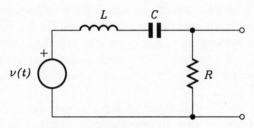

Figure 4.17

4.4 Draw a vector diagram appropriate to finding the resultant of N sinusoidal voltage sources acting together when the amplitude of each source is v and the phases are in arithmetic progression, that is, the phase difference of the nth relative to the first is $(n-1)\delta$, where δ is the phase difference between successive voltages. Show trigonometrically that the resultant amplitude

$$V = v\frac{\sin N\delta/2}{\sin\delta/2}$$

What is the condition on the phase difference δ that the resultant amplitude should be zero?

4.5 Write down the system differential equations for the network shown in Figure 4.18 which includes a unity-coupled transformer with $L_1 = L_2$. Show that for $v(t) = \sin\omega t$ the current $i_2 = a\cos(\omega t - \delta)$ where

$$a = \frac{\omega L}{R}\frac{1}{\sqrt{R^2 + (2\omega L)^2}} \quad \text{and} \quad \tan\delta = \frac{2\omega L}{R}$$

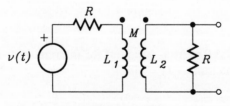

Figure 4.18

5

Network Analysis using Phasors

5.1 Complex Exponential Representation of Simple Harmonic Motion

The elaboration required in the use of trigonometric quantities in regard to the description of oscillations, their combination and their use as forcing functions requiring the solution of differential equations dictates that we adopt another method for representing and combining simple harmonic motion (s.h.m.) using a two-dimensional algebra which will then allow us to bypass the detailed solution of differential equations. The algebra is that of *complex numbers*. The specification of such numbers as points on a two-dimensional surface is developed in the following way. If we envisage positive numbers as steps along a line from left to right, to a set of equally spaced points starting at an arbitrary origin (Figure 5.1), then negative numbers would be steps to equally spaced points in the opposite direction. The positive direction is therefore related to the negative direction by a rotation of 180° and so we may identify negation with a rotation thus:

$$\times(-1) \equiv \text{rotation by } 180°$$

A rotation by 180° is a rotation of 90° performed twice, and so to obtain the multiplication operator corresponding to a rotation of 90° we need to identify that quantity which, when multiplied by itself, produces -1. Clearly

$$\times\sqrt{(-1)} \equiv \text{rotation by } 90°$$

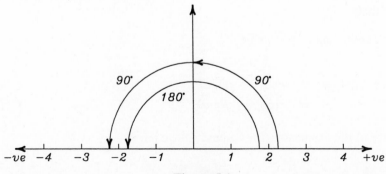

Figure 5.1

63

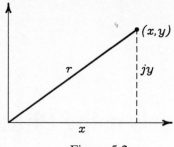

Figure 5.2

and in this way we can define an axis at right angles to the positive axis. A complex number is defined as a point in the plane of the two axes and expressed thus:

$$z = x \times (+1) + y \times \sqrt{(-1)}$$

where z is the complex number and x and y are the displacements along the two perpendicular directions each identified by the appropriate multiplication operators. The symbol $\sqrt{(-1)}$ is considered too cumbersome and replaced by j so that we write:

$$z = x + jy$$

The diagram we have produced is the *Argand diagram* (Figure 5.2), the axes of which are described as *real* $(\times(+1))$ and *imaginary* $(\times\sqrt{(-1)})$. The latter is generally acknowledged to be an unfortunate description which is now too well-established to change.

We now define the *modulus r* of the complex number as the distance from the origin of the point representing the number.

$$r = \sqrt{x^2 + y^2}$$

The *argument* θ is the angle between the line from the origin to the point (x, y) and the real axis. Clearly:

$$\tan\theta = \frac{y}{x}$$

It follows that:

$$x = r\cos\theta \quad \text{and} \quad y = r\sin\theta$$

so that
$$
\begin{aligned}
z = x + jy &= r\cos\theta + jr\sin\theta & (5.1a)\\
&= r(\cos\theta + j\sin\theta) & (5.1b)
\end{aligned}
$$

Making use of the expansions for cos and sin we have:

$$z = r\left\{ \left(1 - \frac{\theta^2}{2!} + \frac{\theta^4}{4!} - \dots\right) + j\left(\theta - \frac{\theta^3}{3!} + \frac{\theta^5}{5!} - \dots\right)\right\}$$

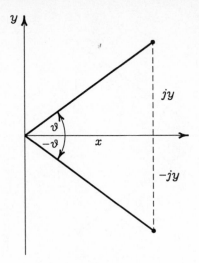

Figure 5.3

Bearing in mind that:

$$j^2 = -1; \; j^3 = -j; \; j^4 = +1; \; j^5 = +j \text{ etc.}$$

we may combine the two series above into one:

$$z \;=\; r\left\{1 + j\theta + \frac{(j\theta)^2}{2!} + \frac{(j\theta)^3}{3!} + \frac{(j\theta)^4}{4!} + \cdots\right\}$$

$$\;=\; re^{j\theta} \quad \text{or} \quad r\exp j\theta$$

This is the Euler form, or the polar form, of a complex number. It was first obtained by Leonhard Euler in 1748 and has often been described as the single most important mathematical expression most engineers and scientists will ever need. By the end of this book at least the ubiquity of the expression should be clear.

The *complex conjugate* of a complex number is obtained by negating the imaginary part of z, which is to take its reflection in the real axis. By reference to Figure 5.3 it will be clear that conjugation also negates the sign of the argument. The following are therefore all equivalent forms of the complex conjugate:

$$z^* \;=\; x - jy$$

$$\;=\; r(\cos\theta - j\sin\theta)$$

$$\;=\; r\exp - j\theta$$

The student should be able to translate freely between the cartesian, trigonometric and polar forms of both a complex number and its conjugate.

Now we return to the use of complex numbers in the representation of s.h.m. We identify the line connecting the origin to the complex number point as the rotating line in our earlier representation. The modulus of the complex number therefore becomes the amplitude of the oscillation and the argument is taken as the phase angle. While we may choose to represent amplitude and phase angle in this way we

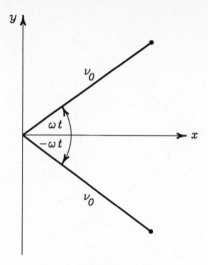

Figure 5.4

cannot alter the fact that a displacement on an Argand diagram and a vector do not share all the same properties. For instance, the product of two complex numbers is always another complex number, while the product of two vectors may be a vector or a scalar. Therefore, somewhat pedantically, we distinguish the new representation from the earlier version by describing the rotating line in the Argand diagram as a rotating *phasor*, and the representation as a *phasor diagram*.

In our new description involving the complex exponential we soon realize that we have more information than we require. In cartesian or trigonometric form we have both real *and* imaginary parts (equation (5.1)). The latter is superfluous given that *any real physical system (such as an electrical network) must be represented by a wholly real number*. The matter may be resolved in a number of ways ranging from the rigorous to the casual. In the most rigorous approach we require that the combination of *two* phasors is necessary, one the conjugate of the other (Figure 5.4). Then:

$$v(t) = \frac{v_0}{2}\{\exp j\omega t + \exp -j\omega t\} \qquad (5.2)$$

$$= \frac{v_0}{2}\{(\cos\omega t + j\sin\omega t) + (\cos\omega t - j\sin\omega t)\}$$

$$= v_0\cos\omega t$$

as required. It may sometimes be necessary to fall back on this approach in particularly difficult problems. It is usually just as satisfactory to stipulate that only the real part is required as follows:

$$v(t) = \text{Re }(v_0\exp j\omega t)$$

In the most casual approach, where great care is needed, it is simply *understood* that only the real part is to be used and we write:

$$v(t) = v_0\exp j\omega t \quad \text{(real part understood)}$$

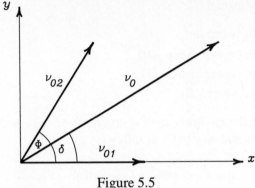

Figure 5.5

In the case of a non-zero initial phase we would write:

$$v(t) = v_0 \exp j(\omega t + \Phi)$$

It is straightforward to factorize this and so separate the time dependence:

$$v(t) \;=\; v_0 \exp j\Phi \exp j\omega t$$
$$\text{or} \quad v(t) \;=\; \mathbf{v} \exp j\omega t$$

where $\mathbf{v} = v_0 \exp j\Phi$ is a complex number (as indicated by the heavy type), stripped of any time dependence but carrying information about both amplitude and initial phase. By convention, the term 'phasor' is always regarded as referring to this complex quantity. It therefore describes the rotating line on the Argand diagram at the instant $t = 0$. It is also sometimes described as the 'complex amplitude', a term more popular outside of engineering.

In solving problems using phasors we combine them by the rules of complex algebra, obtaining the final result by reassigning the overall time dependence and *then* taking the real part. Both real and imaginary parts of the phasors are preserved throughout any solution so that the initial phases are retained as the arguments of the complex exponentials.

Example 5.1 To illustrate the use of the method let us find the resultant of $v_1(t) = $ Re $(v_{01} \exp(j\omega t))$ and $v_2(t) = $ Re $(v_{02} \exp(j\omega t + \Phi))$ which may be shown on a phasor diagram as in Figure 5.5. The phasors are:

$$\mathbf{v}_1 = v_{01} \quad \text{and} \quad \mathbf{v}_2 = v_{02} \exp j\Phi$$

and the resultant phasor is:

$$\mathbf{v} = v_0 \exp j\delta \;=\; \mathbf{v}_1 + \mathbf{v}_2$$
$$=\; v_{01} + v_{02} \exp j\Phi$$
$$=\; v_{01} + v_{02}(\cos\Phi + j\sin\Phi)$$
$$=\; (v_{01} + v_{02}\cos\Phi) + j(v_{02}\sin\Phi)$$

The resultant at any instant is:

$$
\begin{aligned}
v_r(t) &= \text{Re}\,(\mathbf{v}\exp j\omega t) \\
&= \text{Re}\,(v_0\exp j\delta\exp j\omega t) \\
&= \text{Re}\,(v_0\exp j(\omega t + \delta)) \\
&= v_0\cos(\omega t + \delta)
\end{aligned}
$$

where v_0 and δ are the modulus and the argument of the resultant phasor respectively, found in the usual way to be as follows:

$$
\begin{aligned}
v_0 &= \sqrt{(v_{01} + v_{02}\cos\Phi)^2 + (v_{02}\sin\Phi)^2} \\
\tan\delta &= \frac{v_{02}\sin\Phi}{v_{01} + v_{02}\cos\Phi}
\end{aligned}
$$

The result is identical to that obtained by the trigonometric method but the advantage offered is that the manipulation of complex exponentials is so much simpler than the manipulation of trigonometric identities.

5.2   A General Solution for the Free Response

While the system differential equation is simple enough to integrate directly in the case of the RC circuit, it also provides a good opportunity to introduce the more general method and so to compare results. The function e^{st} will solve a linear homogeneous differential equation of any order because of the way in which the exponential function differentiates, reproducing itself. The constant s is to be determined in terms of system parameters. Adopting this approach to the solution of equation (4.4) we write:

$$
y = e^{st} \quad \text{so that} \quad \frac{dy}{dt} = se^{st}
$$

and substituting into equation (4.4) we obtain:

$$
\begin{aligned}
sCRe^{st} + e^{st} &= 0 \\
\text{or} \quad (sCR+1)e^{st} &= 0
\end{aligned}
$$

The solution exists either if $e^{st} = 0$, which is trivial, or if

$$
sCR + 1 = 0 \tag{5.3}
$$

So
$$
s = -\frac{1}{CR} \tag{5.4}
$$

and
$$
y = e^{-\frac{1}{CR}t}
$$

which is the same as the direct solution except that only unit initial conditions have been supposed. The introduction of an amplitude factor v_0 would clearly not affect the solution but has been omitted above to expose the simplicity of the general method. Equation (5.3) is known as the *characteristic* or *auxiliary equation* of the problem and the solution is the *complementary function*. The solution of the homogeneous equation in this way is the first step in obtaining the general solution of a non-homogeneous equation. Finally, we might note that the value of s is *wholly real* in the case of the RC circuit.

In the same way as for the RC circuit we now employ the general method of solution using the exponential function e^{st} for the LC circuit obtaining, from equation (4.12), the characteristic equation:

$$s^2 LC + 1 = 0$$

so that $$s^2 = \frac{-1}{LC} = (j\omega_0)^2$$

where $\omega_0^2 = 1/LC$. Consequently

$$s = \pm j\omega_0 \tag{5.5}$$

and both signs have to be considered in taking the square root. We note that the value of s is *wholly imaginary* in the case of the LC circuit. It is interesting to see how the suggestion, made earlier, that we should use complex quantities (phasors) to represent harmonic oscillations and that they would be needed in conjugate pairs, now emerges from the mathematics in a natural way.

The general solution in this case is obtained by using the trial function with each of the two possibilities for s and taking a linear combination:

$$y = a\exp j\omega_0 t + b\exp - j\omega_0 t \tag{5.6}$$

To satisfy the requirement that the response of a real physical system must be represented by a wholly real number and that at $t = 0$, $y = 1$ we must take $a = b = \frac{1}{2}$, then $y = \cos \omega_0 t$.

It will be useful for future reference to give a summary of the above results in diagrammatic form by recording on an Argand diagram the values of s required in the trial functions in the two cases above. The particular Argand diagram obtained, Figure 5.6, is said to be a representation in the *s-plane*. It is our first contact with a representation which finds wide application throughout network analysis.

5.3   The Complex Frequency

Perversely, before we can pursue our declared aim of utilizing short-hand methods of solution of differential equations we need first to broaden our understanding of the fundamental quantity, *frequency*. We have just seen when examining simple RC and LC circuits that the free response could be described by values of s in e^{st} of either $-1/CR$ (wholly real) or $\pm j/\sqrt{(LC)}$ (wholly imaginary). For circuits of greater complexity we must in general expect the free response to be described by values of s with *both* real and imaginary parts:

$$s = \sigma + j\omega \tag{5.7}$$

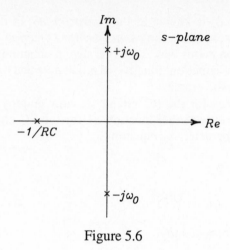

Figure 5.6

which we describe as the *complex frequency*. Points in the s-plane (Figure 5.6) representing the behaviour of such circuits would therefore not be confined to the real and imaginary axes. In this formalism we write an oscillatory voltage, say, as:

$$
\begin{aligned}
v(t) &= v_0 \exp st \quad \text{(real part understood)} \\
&= v_0 \exp (\sigma + j\omega)t \\
&= v_0 \exp \sigma t \exp j\omega t
\end{aligned}
$$

On taking the real part, when the exponential modifier $e^{\sigma t}$ remains unchanged, we obtain:

$$
v(t) = v_0 e^{\sigma t} \cos \omega t
$$

A complex frequency therefore describes an oscillation with an amplitude which either increases exponentially (σ positive), decreases exponentially (σ negative) or remains constant ($\sigma = 0$). The three cases are shown in Figure 5.7. We must eliminate the case of σ positive from further consideration since amplitudes which increase without limit will frustrate solutions for all circuits. The allowed values of s are therefore confined to the left-hand half of the s-plane. Values on the imaginary axis are allowed if they do not correspond to a repeated root. In that case the need for two independent coefficients would require the complementary function to be of the form:

$$
y_c(t) = (K_1 + K_2 t) e^{st}
$$

Such a solution describes a steadily growing response and must be ruled out.

5.4   Response to a Complex Exponential Input

Having established that the complex exponential e^{st}, with s the complex frequency, provides a generalized description of an excitation, it would seem appropriate to examine the response of various circuits to it. The application of the method of

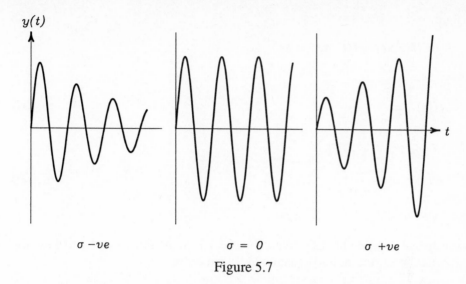

Figure 5.7

undetermined coefficients, in this case to find the particular integral, is almost as straightforward as finding the solution of a homogeneous equation using the exponential function, because of the way in which the exponential differentiates. We therefore apply as input:

$$x(t) = \mathbf{x}e^{st} \qquad (5.8)$$

and write the particular integral:

$$y_p(t) = H(s)\mathbf{x}e^{st} \qquad (5.9)$$

where $H(s)$ is a function of the complex variable s and $\mathbf{x}$ is the phasor for the input. In relating the output to the input, $H(s)$ is that most important quantity which we referred to in Chapter 1 as the *system function* or now, as we are dealing with networks, the *network function*. Expressing the output in terms of its own phasor **y**:

$$y_p(t) = \mathbf{y}e^{st}$$

it follows from equation (5.9) that:

$$\mathbf{y} = H(s)\mathbf{x}$$

The network function therefore relates the input and output phasors, consistent with it being a function of s but not of time.

The two simple circuits we have examined will again serve to illustrate the method.

Example 5.2 For the RC circuit (Figure 4.3), on substituting equations (5.8) and (5.9) into equation (4.3) and using:

$$\frac{dy_p}{dt} = sH(s)\mathbf{x}e^{st}$$

we find:

$$sRCH(s)\mathbf{x}e^{st} + H(s)\mathbf{x}e^{st} = \mathbf{x}e^{st}$$

Hence

$$H(s) \quad = \quad \frac{1}{sCR+1} \tag{5.10a}$$

$$= \quad \frac{1/CR}{s+1/CR}$$

$$\text{or} \quad H(s) \quad = \quad \frac{\alpha}{s+\alpha} \tag{5.10b}$$

where $\alpha = 1/CR$.

Example 5.3 Treating the LC circuit (Figure 4.6) in the same way by making the appropriate substitutions into equation (4.9) and using

$$\frac{d^2y}{dt^2} = s^2 H(s)\mathbf{x}e^{st}$$

we obtain:

$$s^2 LCH(s)\mathbf{x}e^{st} + H(s)\mathbf{x}e^{st} = \mathbf{x}e^{st}$$

Hence

$$H(s) \quad = \quad \frac{1}{s^2 LC+1} \tag{5.11a}$$

$$= \quad \frac{1/LC}{s^2+1/LC}$$

$$\text{or} \quad H(s) \quad = \quad \frac{\omega_0^2}{s^2+\omega_0^2} \tag{5.11b}$$

where $\omega_0^2 = 1/LC$.

In each case the denominator of the network function $H(s)$ is a polynomial in s which follows directly from the differential terms in $y(t)$. Indeed, the denominator is the characteristic equation obtained in the solution of the homogeneous equation. The order of the denominator polynomial therefore reflects the order of the circuit. For the RC circuit (equation (5.10)), $H(s)$ will become infinite ('blow up') when the denominator is zero and this will occur for a value of $s = -1/CR$. The same will happen to $H(s)$ for the LC circuit when $s = \pm j\omega_0$ with $\omega_0 = 1/\sqrt{(LC)}$. The values of s at which $H(s)$, for any network, is infinite are said to be the *poles* of $H(s)$. We may notice immediately that the pole of $H(s)$ for the RC circuit corresponds to the value of s which describes the free response of that circuit (equation (5.4)). An inspection of equation (5.5) reveals that a similar statement is true for the poles of $H(s)$ for the LC circuit. What we have previously described simply as a representation in the s-plane of the free response of two circuits (Figure 5.6) now takes on an added significance: it is a *pole diagram*. It should be intuitively obvious that there would be such a correspondence between the poles and the free response. We should expect that when any system is stimulated in a way corresponding to the way in which it behaves naturally then the response will be anomalous: either indeterminate or at least very high.

When depicting the free response of networks by points in the *s*-plane it was felt necessary to exclude points in the right-hand half-plane as these would represent oscillations growing without limit. It follows that when we now consider systems characterized by the poles in their network functions we must confine ourselves to such systems as have poles in the left-hand half of the *s*-plane. A system with a pole in the right-hand half of the *s*-plane would possess an exponentially growing free response which would inevitably be excited and mask all other behaviour.

5.5   The *n*th Order System

So far we have dealt only with first-order and second-order systems and then with simple examples of these. Such an approach can be justified on the grounds that in practice many high-order systems are realized by cascading such elementary building blocks. However, we must in general expect to be able to cope with systems of arbitrary order which are not obviously the combination of first-order and second-order units. The starting point for such an analysis is the general system differential equation, equation (5.12), which expresses the possibility that not only does the forcing function equate to a linear combination of the output and its derivatives, but also that it, in turn, may be given by a linear combination of the input and its derivatives:

$$a_n\frac{d^n y}{dt^n} + \ldots + a_1\frac{dy}{dt} + a_0 y = b_m\frac{d^m x}{dt^m} + \ldots + b_1\frac{dx}{dt} + b_0 x \qquad (5.12)$$

Again considering the behaviour for the complex exponential input, the particular integral, the steady state response after the effect of the initial conditions has died away, is as given in equation (5.9) with the input as in equation (5.8). Making these substitutions in equation (5.12):

$$(a_n s^n + \ldots + a_1 s + a_0)H(s)\mathbf{x}e^{st} = (b_m s^m + \ldots + b_1 s + b_0)\mathbf{x}e^{st}$$

Dividing throughout by $\mathbf{x}e^{st}$ and rearranging we obtain:

$$H(s) = \frac{b_m s^m + \ldots + b_1 s + b_0}{a_n s^n + \ldots + a_1 s + a_0} \qquad (5.13)$$

In general, therefore, the network function $H(s)$ is the quotient of two polynomials. Each may be expressed as a product of factors involving their respective roots:

$$H(s) = \frac{(s - z_1)(s - z_2)\ldots(s - z_m)}{(s - p_1)(s - p_2)\ldots(s - p_n)}$$

We have already encountered the idea of the roots of the denominator polynomial, $p_1, \ldots, p_n$, as poles of $H(s)$. In a similar way, the roots of the numerator polynomial, $z_1, \ldots, z_m$, correspond to *zeros* in $H(s)$. Both may be plotted in the *s*-plane producing a *pole-zero* diagram (Figure 5.8). Any particular value of *s* for which we

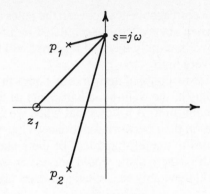

Figure 5.8

require $H(s)$ may be represented by a point in the s-plane. Typically this will be a point on the imaginary axis representing a steady a.c. input (the subject of Chapter 8). The factors of both the denominator and the numerator may be seen as displacements from the chosen point to each of the poles and zeros. Conversely, $H(s)$ may be regarded as a ratio of the products of vectors in the s-plane.

5.6 The Complex Impedance

It was stated earlier that our objective was the elimination of direct reference to system differential equations. A less than completely general method of achieving this can be developed by considering the response of the constituent components, R, L and C respectively, to a complex exponential input. Recalling from Chapter 1 the fundamental current–voltage relationships:

$$v(t) = Ri(t), \quad v(t) = L\frac{di}{dt}, \quad i(t) = \frac{dq}{dt} = \frac{d(Cv)}{dt}$$

we apply in each case a complex exponential voltage source, $v(t) = \mathbf{v}e^{st}$, and then find the current, itself also described by a complex exponential, $i(t) = \mathbf{i}e^{st}$ ($\mathbf{i}$ and $\mathbf{v}$ are the current and voltage phasors). The scheme will in fact need little modification to render it more general in application. For a resistor (Figure 5.9(a)):

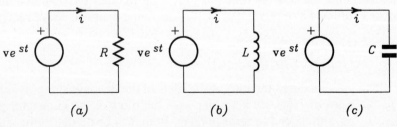

Figure 5.9

$$\mathbf{v}e^{st} = R\mathbf{i}e^{st}$$

so that $\quad \mathbf{v} = R\mathbf{i}$ $\hfill$ (5.14)

For an inductor (Figure 5.9(b)):

$$\mathbf{v}e^{st} = L\frac{d}{dt}\mathbf{i}e^{st}$$
$$= sL\mathbf{i}e^{st}$$

so that $\quad \mathbf{v} = sL\mathbf{i}$ $\hfill$ (5.15)

and for a capacitor (Figure 5.9(c)):

$$\mathbf{i}e^{st} = C\frac{d}{dt}\mathbf{v}e^{st}$$
$$= sC\mathbf{v}e^{st}$$

so that $\quad \mathbf{v} = \frac{1}{sC}\mathbf{i}$ $\hfill$ (5.16)

Equation (5.14) is the form of Ohm's law appropriate to time-varying quantities now written for an instant in time, $t = 0$, in terms of the current and voltage phasors. Equations (5.15) and (5.16) are analogous statements for inductance and capacitance. In the same way that we could relate voltages and currents in individual resistors using Ohm's law when dealing with d.c. sources, we can now use equations (5.14), (5.15) and (5.16) when dealing with networks containing R, L and C energized by a complex exponential source. The various network equations and theorems which are applicable at all times to networks containing only resistors and d.c. sources, apply equally well at an instant in time to those containing R, L and C and time-dependent sources. It follows that the relationship between the current and voltage phasors for a combination of components may be evaluated using the same rules as for resistors. The simple existence of analogous relationships between current and voltage, *at any instant in time*, is taken as sufficient justification for this proposition. However, one major point to note is that the quantity connecting current and voltage is now, in general, complex. For this reason the term *impedance* and the symbol Z are introduced to identify it, and R, sL and $1/sC$ are all said to be contributions to the complex impedance.

Example 5.4 To illustrate the role of the impedance in obtaining the network function without recourse to the system differential equation, consider once again the simple RC circuit, Figure 5.10(a). We can regard this now as an impedance potential divider (Figure 5.10(b)) and could write down the ratio of the output to the input by inspection. However, on this one occasion we will show the full development starting from Kirchhoff's laws. Kirchhoff's voltage law written in terms of the phasors, currents and voltages at the instant $t = 0$ yields:

$$\mathbf{x} = R\mathbf{i} + \frac{1}{sC}\mathbf{i}$$

with $\quad \mathbf{y} = \frac{1}{sC}\mathbf{i}$

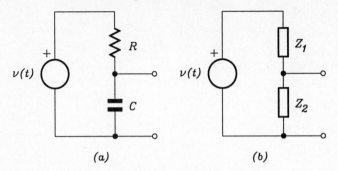

(a)   (b)

Figure 5.10

so

$$\frac{\mathbf{y}}{\mathbf{x}} = H(s) \quad = \quad \frac{1/sC}{R+1/sC} \tag{5.17a}$$

$$= \quad \frac{1}{sCR+1} \tag{5.17b}$$

Treating the circuit as a potential divider we could have obtained equation (5.17a) directly.

Example 5.5 As a further example we may treat the LC circuit again but now with added resistance to make it more realistic (Figure 5.11). The impedance potential divider approach now yields:

$$H(s) \quad = \quad \frac{R}{R+1/sC+sL}$$

$$= \quad \frac{sR/L}{s^2+sR/L+1/LC}$$

$$= \quad \frac{s\omega_0/Q}{s^2+s\omega_0/Q+\omega_0^2}$$

where $\omega_0^2 = 1/LC$ and $Q = 1/\omega_0 CR$. The full significance of the latter quantity, the 'quality factor', will not become apparent until Chapter 8.

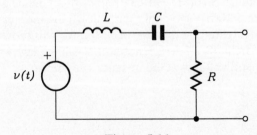

Figure 5.11

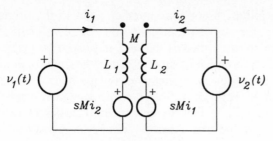

Figure 5.12

5.7 Networks with Transformers

We now examine the situation where the network includes a transformer (Figure 4.15) and the sources are the complex exponential functions $v_1(t) = \mathbf{v}_1 e^{st}$ and $v_2(t) = \mathbf{v}_2 e^{st}$. The corresponding currents will be complex exponentials distinguished by their own phasors. Dividing throughout by e^{st}, following substitution into equations (4.25a) and (4.25b), produces:

$$\mathbf{v}_1 = sL_1\mathbf{i}_1 + sM\mathbf{i}_2 \tag{5.18a}$$
$$\mathbf{v}_2 = sM\mathbf{i}_1 + sL_2\mathbf{i}_2 \tag{5.18b}$$

From this it will be seen that the effect of the coupling may be represented by the addition of an extra source in each of the circuits (Figure 5.12) with the coupling otherwise reduced to zero. In a situation where $v_2(t)$ exactly balances the open circuit voltage of the secondary reducing the secondary current to zero, equations (5.18a) and (5.18b) reduce to:

$$\mathbf{v}_1 = sL_1\mathbf{i}_1 \quad \text{and} \quad \mathbf{v}_2 = sM\mathbf{i}_1$$

The source in the secondary is the same as the open-circuit voltage of the secondary except for a phase difference of π, the voltage transfer function is therefore given by the ratio:

$$H(s) = -\frac{\mathbf{v}_2}{\mathbf{v}_1} = -\frac{M}{L_1}$$

For the case of a unity-coupled $(k = 1)$ transformer we have, using equation (4.23):

$$H(s) = -\frac{N_2}{N_1} = -n$$

which gives a wider interpretation to the turns ratio than we obtained in the example in Chapter 4.

Summary

The rotating vector representation of simple harmonic motion introduced earlier, and the associated trigonometric manipulation required in solutions, is laborious. An alternative approach is available using the algebra of *complex numbers* where we utilize the line joining a complex number point to the origin in the *Argand diagram* as the basis for our new representation. Displacements along the axes of the Argand diagram are described as the *real* and *imaginary* parts of the complex number in its cartesian form. Much more useful is the *Euler* or *polar* representation where we identify the angle (the *argument* of the number) with the angle changing with time in the earlier vector diagram. The time dependence specified in this way is therefore given by $e^{j\omega t}$. In fact the new method carries more information than necessary. Bearing in mind that *a real physical system must be represented by a wholly real number* we must restrict ourselves to using only the *real part* of the complex number, which is $\cos \omega t$ as in the use of the vector diagram. There are rigorous and more casual ways of implementing this restriction.

Non-zero initial phase for periodic voltages and currents may be described by an angle, Φ, included in the argument of the complex number thus: $\exp(j\omega + \Phi)$. Then for an oscillating voltage, say, of amplitude v_0 the factor $v_0 \exp j\Phi$, stripped of the time dependence $\exp j\omega t$, is described as the *phasor* and the Argand diagram in this application as a *phasor diagram*, so avoiding confusion with the earlier vector diagram representation of s.h.m.

Returning to the solution of the differential equations for simple networks we employed a general method of solution involving the exponential function e^{st} (more elaborate than strictly necessary in such simple cases) and found that s may in some cases be real and in others imaginary. The general expectation is that s will be complex, that is, have both real *and* imaginary parts. Such a factor, $s = \sigma + j\omega$, having the dimensions $\sec^{-1}$, is termed the *complex frequency* and can simultaneously describe a steady change (increase or decrease) in amplitude impressed upon a time-dependent fluctuation.

The complex exponential e^{st} appears to provide such a general description of a disturbance that it is instructive to use it to describe voltages and currents exciting electronic circuits. In so doing and solving the system differential equations we obtain functions of the complex frequency $H(s)$ which relate output to input. They are the system functions, or in this case the *transfer functions* for the networks considered. The denominator of $H(s)$ is seen to reflect the order of the system in the highest power of s. The nth order system will have a term in s^n in the denominator.

Applying sources where the time dependence is given by a complex exponential to the simple components resistor, capacitor and inductor, a means of relating voltage to current for a capacitor and for an inductor in a way analogous to that for a resistor (Ohm's law) was developed, so defining *impedances*. The impedance of a capacitor is $1/sC$, while that for an inductor is sL; that for a resistor is simply its resistance, R.

The idea of an impedance potential divider becomes a useful concept allowing methods developed for purely resistive networks to be applied to those now containing capacitors and/or inductors in addition to resistors. Impedances employed in this way allow us to bypass any reference to system differential equations.

The complex exponential method was also useful in treating networks with mutual inductance (transformers) where the turns ratio for a unity-coupled transformer was seen to take on the role of the modulus of the system function.

Problems

5.1 N sinusoidal voltage sources each have amplitude v and their phases are in arithmetic progression, that is, the phase difference of the nth relative to the first is $(n-1)\delta$, where δ is the phase difference between successive voltages. Show by the use of complex exponentials that the resultant amplitude of the sources acting together is:

$$V = v\frac{\sin N\delta/2}{\sin \delta/2}$$

5.2 Use the complex exponential e^{st} as a trial solution to find the free response of the network shown in Figure 5.11 for the case $4/LC > R^2/L^2$. (The system differential equation for this network was found in Problem 4.3.) Show schematically on an Argand diagram the values of s required for the solution.

5.3 Obtain the system differential equations for each of the networks shown in Figure 5.13. Find the ratio of output to input (the system transfer functions) when the complex exponential e^{st} is applied as input.

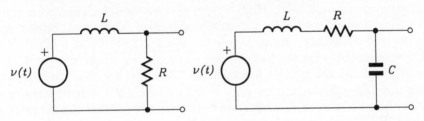

Figure 5.13

5.4 By first representing each component in terms of its impedance, obtain the voltage transfer function for each of the networks in Figure 5.13.

5.5 Find the system transfer function for the network in Figure 5.14 which includes a unity-coupled transformer with $L_1 = L_2$.

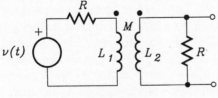

Figure 5.14

6

The Laplace Transform in Network Analysis

6.1 Operational Calculus

We saw in Chapter 4 how the solution of the system differential equation for the first-order system with the simplest inputs required considerable effort even when specialized methods were used. Higher-order systems would require correspondingly greater effort. The need for a streamlined method is apparent. Such a need was recognized by, among others, Oliver Heaviside (1850–1925), a largely self-taught, eccentric but brilliant physicist and engineer. Heaviside's approach was the use of an operational calculus in which the differential operator is treated symbolically and differential equations are solved by essentially algebraic techniques. Operational methods are in fact common in science and engineering but often go unrecognized. The computer language BASIC is an operational method for instructing a computer, an alternative to the daunting prospect of communicating in binary code. The compiler (or interpreter) transforms the source code to machine instructions. Using logarithms is an operational method for performing multiplication and division but this is a statement which becomes less well understood by students with every new generation of hand-held calculators.

Heaviside's methods were heuristic and lacked rigour but have since been shown to be without error. However, as his contemporaries were quick to point out, an alternative operational method, derived by Laplace, had already been in existence for some fifty years. Laplace had certainly not had network problems principally in mind when developing his techniques, if he had given them any thought at all. It was never clear if Heaviside was aware of the use of the Laplace transform or just chose to ignore it. His own method can also be said to have been presaged by Cauchy even earlier. It is a great pity that so much has been made of the controversy surrounding Heaviside's work because, quite apart from his method, he made a major contribution and provided a great impetus to the subject of circuit analysis. The fact is, however, that the operational method now in universal use is that derived by Laplace. The most likely explanation of this is that, while both methods achieve the same results, Heaviside's method does require the differential equation as its starting point. We can entirely avoid having to write down differential equations by using Laplace transforms and transforming the equations of the constituent components to obtain their impedances. We will then be able to write down the network function in the same way as we could for the complex exponential input e^{st}, indeed it is the same function when there is *no initial stored energy* in the circuit.

6.2 The Laplace Transform

Recall from Chapter 4 that we could solve the first-order non-homogeneous differential equation:

$$\frac{dy(t)}{dt} + ay(t) = x(t)$$

by multiplying throughout by e^{at} which enabled us to write it in the form, equation (4.15):

$$\frac{d}{dt}(e^{at}y(t)) = e^{at}x(t)$$

Integration then yielded equation (4.16):

$$y(t) = ce^{-at} + e^{-at}\int e^{at}x(t)dt$$

An alternative approach, rather than to try to proceed directly to the solution, is to multiply throughout by a factor, called the *kernel*, and integrate to produce in place of each original term its transform, a function of s. For a function of time $f(t)$, a kernel of the form e^{-st} produces the Laplace transform $F(s)$:

$$F(s) = \int_{0-}^{\infty} f(t)e^{-st}dt \tag{6.1}$$

The integration is taken from $0-$ to ∞ to include any discontinuities occurring at $t = 0$.

We saw in Chapter 5 that we should in general describe the coefficient of time in an exponential as a *complex frequency*. The Laplace transform therefore produces a function in the complex frequency domain, the s-plane, from a time domain function.

The operational role of the Laplace transform may be seen by considering its application in the first instance to $df(t)/dt$. Representing the transform by $\mathcal{L}$ we have:

$$\mathcal{L}\left[\frac{df(t)}{dt}\right] = \int_{0-}^{\infty}\frac{df(t)}{dt}e^{-st}dt$$

Integrating by parts we may readily show that:

$$\int_{0-}^{\infty}\frac{df(t)}{dt}e^{-st}dt = s\int_{0-}^{\infty}f(t)e^{-st}dt - f(0-)$$

where $f(0-)$ is $f(t)$ evaluated at an infinitesimal instant of time before $t = 0$. Introducing the Laplace transform of $f(t)$ from equation (6.1) it follows that:

$$L\left[\frac{df(t)}{dt}\right] = sF(s) - f(0-) \tag{6.2}$$

an *algebraic* expression in place of the original *differential* expression.

The reader may like to show that the transform of the second differential coefficient of a function:

$$L\left[\frac{d^2f(t)}{dt^2}\right] = s^2F(s) - sf(0-) - \frac{df(0-)}{dt} \tag{6.3}$$

where $df(0-)/dt$ is df/dt evaluated at an infinitesimal instant before $t = 0$.

Another important property of the Laplace transform is that the transform of a sum is the sum of the transforms. It follows that Kirchhoff's laws may equally well be stated in terms of the voltages and currents as functions of s as of functions of t. Similarly, the principle of superposition may be applied to Laplace transforms. Indeed, all network laws and theorems may be drafted in terms of Laplace transforms. Other simple rules obtain for the operations of integration, scaling, time-shifting and so on. To avoid obscuring the principles of the method these other operational transforms are consigned to Appendix C, together with associated theorems.

The general expectation is that the application of the transform method to a particular system will produce a function, $F(s)$, describing the s-domain response of the system. Laplace transformation only has value as an operational method by virtue of the fact that the inverse operation, producing $f(t)$ from $F(s)$, does exist and is unique. (For a proof of this the reader is referred to specialized texts.) We are able therefore, by inverse transformation to obtain the time domain response, $f(t)$, of the system.

The Laplace transform of almost any function of interest will usually already have been found and may be referred to in appropriate texts. A selection of transform pairs which will enable us to solve simple problems is given in Table 6.1. In each case there is a one-to-one relationship between the function and its transform. When either is known the other is determined. That is, representing the inverse transform symbolically by L^{-1}:

$$L[f(t)] = F(s) \quad \text{and} \quad L^{-1}[F(s)] = f(t)$$

Occasionally the application of the Laplace transform method to a particular system may result in a function $F(s)$, the inverse transform of which cannot be obtained by reference to known transform pairs. It may then in principle be evaluated using the inversion integral:

$$f(t) = \frac{1}{2\pi j}\int_{\sigma-j\infty}^{\sigma+j\infty} F(s)e^{st}ds \tag{6.4}$$

The requirement is that the integral be performed over the complete range of s but, as the route from $-\infty$ to $+\infty$ is immaterial, we choose a route along constant σ (Figure 6.1(a)). A positive σ ensures that the inverse operation (the forward transform) will contain $e^{-\sigma t}$, an exponential decay, and will therefore converge

$f(t)$	$F(s)$
$\delta(t)$	1
$u(t)$	$\dfrac{1}{s}$
$u(t)e^{-\alpha t}$	$\dfrac{1}{s+\alpha}$
$u(t)\cos\omega t$	$\dfrac{s}{s^2+\omega^2}$
$u(t)\sin\omega t$	$\dfrac{\omega}{s^2+\omega^2}$
$u(t)e^{-\alpha t}\cos\omega t$	$\dfrac{s+\alpha}{(s+\alpha)^2+\omega^2}$
$u(t)e^{-\alpha t}\sin\omega t$	$\dfrac{\omega}{(s+\alpha)^2+\omega^2}$

Table 6.1

(produce a finite result). Were it necessary to evaluate the integral in this form it might be more convenient to express the range of integration in terms of a quantity which approaches infinity:

$$f(t) = \frac{1}{2\pi j}\lim_{R\to\infty}\int_{\sigma-jR}^{\sigma+jR}F(s)e^{st}ds$$

In fact any $F(s)$ of interest should decrease at least as fast as $1/s$ as s increases. It may otherwise be taken as the sum of factors which so decrease (partial-fraction expansion is dealt with shortly). It follows that the integration between limits required for the inverse transform may be performed as a closed path integral, part of which is the arc of a circle of radius R on which the contribution to the integral approaches zero as $R\to\infty$:

$$f(t) = \frac{1}{2\pi j}\lim_{R\to\infty}\oint F(s)e^{st}ds$$

where it is understood that the path followed is in a counter-clockwise sense around the left-hand s-plane (Figure 6.1(b)). It is only along this route that one may guarantee the extra contribution to be zero for $t > 0$. By the application principally of the residue theorem relating to integration in the complex plane, one may evaluate inverse transformations where no tabulated solutions exist. Here is another major area of mathematics which makes an impact on the subject of network analysis. The reader should again consult specialized texts for more on contour integration in the complex plane. The simple tabulated solutions will be adequate for our purpose.

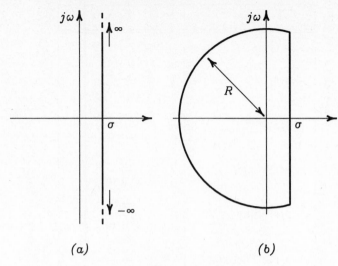

Figure 6.1

6.3 Application to Network Equations

We have throughout this text used both the RC and the L C circuits as illustrations of technique. We continue this approach here.

Example 6.1 Applying the Laplace transform method first to equation (4.3) for the RC circuit:

$$CR\frac{dy(t)}{dt} + y(t) = x(t)$$

Multiplying throughout by e^{-st} and integrating:

$$CR\int_{0-}^{\infty}\frac{dy(t)}{dt}^{-st}dt + \int_{0-}^{\infty}y(t)e^{-st}dt = \int_{0-}^{\infty}x(t)e^{-st}dt$$

we have an equation in the Laplace transforms of the original terms. Using equation (6.2) for the transform of the differential coefficient and writing the transforms of $x(t)$ and $y(t)$ as:

$$X(s) = \int_{0-}^{\infty}x(t)e^{-st}dt \quad \text{and} \quad Y(s) = \int_{0-}^{\infty}y(t)e^{-st}dt$$

respectively, the original equation is reduced to:

$$CRsY(s) - CRy(0-) + Y(s) = X(s)$$

an algebraic equation in the variable s. The transform of the output is then:

$$Y(s) = \frac{X(s)}{sCR+1} + \frac{CRy(0-)}{sCR+1}$$

In the case where there is no initial stored energy, $y(0-) = 0$ and we recognize the ratio:

$$\frac{Y(s)}{X(s)} = \frac{1}{sCR+1} \tag{6.5}$$

as the expression for the system function of the RC circuit obtained in Chapter 5, equation (5.10a).

Example 6.2 Turning now to equation (4.9) for the LC circuit:

$$LC\frac{d^2y(t)}{dt^2} + y(t) = x(t)$$

we shall dispense with the step of multiplying by the kernel and integrating and instead proceed directly to the Laplace transform. Using equation (6.3) for the transform of the second differential we obtain from equation (4.9):

$$s^2LCY(s) - sLCy(0-) - LC\frac{dy(0-)}{dt} + Y(s) = X(s)$$

again an algebraic equation in place of a differential equation. So:

$$Y(s) = \frac{X(s)}{s^2LC+1} + \frac{sLCy(0-)}{s^2LC+1} + \frac{LCdy(0-)/dt}{s^2LC+1}$$

On the assumption of zero initial stored energy, $y(0-) = 0$. We may also stipulate that $dy(0-)/dt = 0$. The latter condition follows from the fact that, while the input $x(t)$ may undergo a discontinuous change, neither the current through nor the voltage across any component may change discontinuously. It follows that for an nth order system all derivatives of y up to order $(n-1)$ must be continuous for $t \geq 0$. We therefore find:

$$\frac{Y(s)}{X(s)} = \frac{1}{s^2LC+1} \tag{6.6}$$

and recognize this as the expression for the system function of the LC circuit obtained in Chapter 5, equation (5.11a).

6.4   The General Interpretation of the System Function

From the treatment of the RC and LC circuits we have seen that the ratio of the Laplace transform of the output to the Laplace transform of the input is the same, in the case of zero initial energy, as the function we have previously called the system function, $H(s)$, when treating systems subject to a complex exponential input. The complex exponential input allowed a very specialized solution of the system differential equation whereas the Laplace transform method will allow completely general solutions. We might therefore conclude from this alone that the more general interpretation to be given to the system function is that it is the ratio of Laplace transforms. Indeed, this is the interpretation we wish to promote and now confirm by means of a rudimentary proof where some appeal to intuition will be made.

The inverse Laplace transform, equation (6.4), which yields the input function $x(t)$ is:

$$x(t) = \frac{1}{2\pi j} \int X(s)e^{st}ds$$

(omitting the limits which are not essential to the present argument). An interpretation we may place on the inverse transform is that it is the sum taken over the whole range of complex exponentials, which together constitute the input, in the limit of components infinitesimally separated in s. The amplitude of any component is $X(s)$. In the absence of any initial stored energy we might expect to find the output by scaling each input component by $H(s)$, as in Chapter 5, and again integrate over components:

$$y(t) = \frac{1}{2\pi j} \int H(s)X(s)e^{st}ds \tag{6.7}$$

The output $y(t)$ will, however, also be given directly by an inverse transform:

$$y(t) = \frac{1}{2\pi j} \int Y(s)e^{st}ds \tag{6.8}$$

Comparison of equations (6.7) and (6.8) shows that:

$$Y(s) = H(s)X(s)$$

$$\text{or} \quad H(s) = \frac{Y(s)}{X(s)} \tag{6.9}$$

which confirms the earlier supposition that $H(s)$ should be regarded in general as the ratio of the Laplace transforms of the output and the input. One must, however, make the proviso that this form of $H(s)$ may only be employed to find the response of a system to a general input, subject to the conditions under which it was derived, namely zero initial stored energy. The response so obtained is known as the *zero state response* (ZSR). The response is otherwise complete, that is, it includes both transient and steady state contributions, corresponding to the complementary function and the particular integral of the earlier classical treatment.

6.5   The Laplacian Impedance

We might remark again at this stage that the declared objective of this text is to eliminate the work involved in dealing directly with system differential equations. In Chapter 5 we avoided reference to differential equations by introducing the impedance of circuit elements. We obtained $H(s)$ both by introducing impedances into equations given by the application of Kirchhoff's laws and by the application of the potential-divider principle. It appears that taking the Laplace transform of a differential equation term-by-term leads to a ratio of Laplace transforms which we now identify as the system function. With a view to bypassing this last step, let us transform the equations for each of the constituent components. For a resistor, transforming:

$$v(t) = Ri(t)$$

we obtain:

$$V(s) = RI(s) \quad \text{and the ratio} \quad \frac{V(s)}{I(s)} = R \tag{6.10}$$

while for an inductor

$$v(t) = L\frac{di(t)}{dt}$$

transforms to:

$$V(s) \;=\; sLI(s) - Li(0-) \tag{6.11a}$$

$$\text{or} \quad I(s) \;=\; \frac{V(s)}{sL} + \frac{i(0-)}{s} \tag{6.11b}$$

In the case of zero initial stored energy we obtain:

$$V(s) = sLI(s) \quad \text{and the ratio} \quad \frac{V(s)}{I(s)} = sL \tag{6.12}$$

Lastly, for a capacitor,

$$i(t) = C\frac{dv(t)}{dt}$$

produces:

$$I(s) \;=\; sCV(s) - Cv(0-) \tag{6.13a}$$

$$\text{or} \quad V(s) \;=\; \frac{I(s)}{sC} + \frac{v(0-)}{s} \tag{6.13b}$$

which in the case of zero initial stored energy is:

$$I(s) = sCV(s) \quad \text{and the ratio} \quad \frac{V(s)}{I(s)} = \frac{1}{sC} \tag{6.14}$$

The quantities R, sL and $1/sC$ are here contributions to the *generalized or Laplacian impedance*. Algebraically they are identical to the complex impedances we introduced in the case of the complex exponential input. The new interpretation embraces the former.

Example 6.3 Echoing the procedure of Chapter 5 we once again use the RC circuit (Figure 6.2) to illustrate the use of impedances, but now in their role as the ratio of Laplace transforms. Kirchhoff's voltage law, applied to the voltage drops taken clockwise around the circuit as detailed in section 2.1, requires that:

$$v_R(t) + v_C(t) - x(t) = 0$$

while $\quad y(t) = v_C(t)$

where $x(t)$ is now a generalized input. Taking the Laplace transform throughout:

$$V_R(s) + V_C(s) - X(s) = 0$$

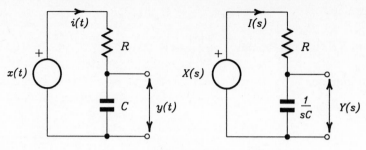

Figure 6.2

while $Y(s) = V_C(s)$

Introducing the current, in the form of its Laplace transform, and using the generalized impedance:

$$RI(s) + \frac{1}{sC}I(s) - X(s) = 0$$

or $X(s) = RI(s) + \frac{1}{sC}I(s)$

and $Y(s) = \frac{1}{sC}I(s)$

lead to:

$$\frac{Y(s)}{X(s)} = H(s) \quad = \quad \frac{1/sC}{R + 1/sC} \tag{6.15a}$$

$$= \quad \frac{1}{sCR + 1}$$

$$= \quad \frac{1/CR}{s + 1/CR} \tag{6.15b}$$

We have obtained equation (6.15b) without reference to the system differential equation which follows from the fact that we have not needed to make reference to the differential expressions for the individual components. Instead we have used for each element its generalized impedance. Treating the circuit as an impedance potential divider, as in Chapter 5, would produce equation (6.15a) directly and that eliminates the step of applying Kirchhoff's laws. The method of complex impedances will therefore usually allow the system function $H(s)$ to be written by inspection. In circuits which are more complicated than a simple potential divider, systematic reduction by the application of Thevenin's theorem, will allow $H(s)$ to be found.

6.6   Finding the Time Domain Response

We are now able using component impedances to write down the system function $H(s)$ for any required system. For very many cases of interest the input $x(t)$ will have a well-known and tabulated transform $X(s)$ so that, when there is no initial stored energy, we may then find the Laplace transform of the output using:

$$Y(s) = H(s)X(s)$$

(we see later how to include stored energy). As was suggested earlier, we expect that the inverse transform of $Y(s)$ will yield the time domain response, $y(t)$, of the system. However, it is unlikely that $Y(s)$ will be in a form that lends itself directly to inverse transformation by reference to tabulated transforms. There are a number of well-tried procedures we may use to render $Y(s)$ amenable to such simple inverse transformation.

The general form of $Y(s)$ will have much in common with the general form of $H(s)$ (Chapter 5), being characterized most importantly by its poles (roots of the denominator):

$$
\begin{aligned}
Y(s) &= \frac{b_m s^m + \ldots + b_1 s + b_0}{a_n s^n + \ldots + a_1 s + a_0} \\
&= \frac{(s - z_1)(s - z_2)\ldots(s - z_m)}{(s - p_1)(s - p_2)\ldots(s - p_n)}
\end{aligned}
\tag{6.16}
$$

A function of s with but a single term in the denominator:

$$Y(s) = \frac{K}{s - p}$$

would be straightforward to inverse transform by reference to tabulated transforms:

$$y(t) = Ku(t)e^{pt}$$

The product of factors in the denominator of the general expression makes such inverse transformation impossible. We require instead a *partial fraction expansion* of $Y(s)$:

$$Y(s) = \frac{K_1}{s - p_1} + \frac{K_2}{s - p_2} + \ldots + \frac{K_n}{s - p_n} \tag{6.17}$$

for each factor of which the inverse transform could be written immediately. Such partial fraction expansion itself has a prerequisite, namely that $Y(s)$ is a *proper fraction*, that is, the highest power of s in the numerator should be less than the highest power of s in the denominator. When that is not the case (we have an *improper fraction*), long division of the numerator by the denominator must be used to produce a proper fraction as the remainder of the division together with terms in s^n.

Example 6.4 $Y(s) = (s^2 - 5s + 6)/(s^2 - 5s + 4)$ is not a proper fraction. How to proceed in this case should be self-evident; but formally, by long division:

$$
\begin{array}{r}
1 \\
s^2 - 5s + 6 \overline{\smash{)}\, s^2 - 5s + 4} \\
\underline{s^2 - 5s + 6} \\
-2
\end{array}
$$

we obtain:

$$Y(s) = 1 - \frac{2}{s^2 - 5s + 4}$$

Another prerequisite is that the coefficient of s in the denominator should be unity. Division of both numerator and denominator by the coefficient of s^n will reduce $Y(s)$ to the required form. When both these requirements have been met a particular coefficient K may be found by multiplying both equations (6.16) and (6.17) by the corresponding denominator and setting s equal to the value which renders that denominator zero (the value of that particular pole of the expression). The above procedure has been cast in a shorthand form, known as the *Cover-up Rule*, implemented thus: cover-up one denominator in the original expression and set s to the value of the corresponding pole throughout the remainder of the expression. The procedure is best illustrated by example.

Example 6.5 Continuing with the RC circuit as a vehicle to understand a new method we find the response to $u(t)$ in the case of zero initial stored energy. The response to $u(t)$ characterizes the behaviour of circuits during the switching operations to which any circuit must be subject. The Laplace transform of $u(t)$ is particularly simple, $1/s$ (Table 6.1). Making the appropriate substitution in equation (6.15b):

$$Y(s) = \frac{1}{s} \frac{1/CR}{s + 1/CR} \tag{6.18}$$

We have here the *product* of two terms the inverse transform of each of which could be obtained from Table 6.1. To effect the inverse transform of equation (6.18) we require a sum of terms obtained by a *partial fraction expansion*:

$$Y(s) = \frac{K_1}{s} + \frac{K_2}{s - p} \quad \text{with} \quad p = -1/CR \tag{6.19}$$

A particular coefficient K is found, as suggested above, by multiplying both equations (6.18) and (6.19) by the corresponding denominator and setting s equal to the value which renders that denominator zero (the value of that particular pole of the expression):

$$sY(s) \quad = \quad \frac{-p}{s - p} = K_1 + \frac{sK_2}{s - p}$$

$$[sY(s)]_{s=0} \quad = \quad 1 = K_1$$

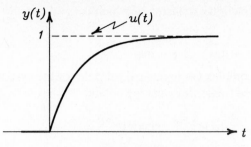

Figure 6.3

Similarly

$$(s-p)Y(s) \;=\; \frac{-p}{s} = \frac{(s-p)K_1}{s} + K_2$$

$$[(s-p)Y(s)]_{s=p} \;=\; -1 = K_2$$

It is important to note that we must multiply through by the denominator and *then* set that denominator to zero.

Implementing the above procedure by the use of the Cover-up Rule:

$$K_1 = \left[\frac{1}{\boxed{s}} \frac{-p}{s-p} \right]_{s=0} = 1$$

$$K_2 = \left[\frac{1}{s} \frac{-p}{\boxed{s-p}} \right]_{s=p} = -1$$

The boxed denominator is the one covered-up. By either method:

$$Y(s) = \frac{1}{s} - \frac{1}{s-p}$$

the inverse transform of which, using Table 6.1, is:

$$y(t) \;=\; u(t) - u(t)e^{pt}$$
$$\;=\; u(t)[1 - e^{pt}]$$

and is shown in Figure 6.3.

Example 6.6 Consider again the somewhat unrealistic (infinite Q) LC circuit dealt with in Example 5.3, for the purpose of demonstrating here partial fraction expansion for a second-order system. The ZSR for the input $u(t)$ may be found using equation 5.11(b),

$$H(s) = \frac{Y(s)}{X(s)} \;=\; \frac{\omega_0^2}{s^2 + \omega_0^2}, \quad \text{with} \quad X(s) = 1/s$$

That is

$$Y(s) \;=\; \frac{1}{s} \frac{\omega_0^2}{s^2 + \omega_0^2}$$

the complete partial fraction expansion of which is:

$$Y(s) = \frac{1}{s} - \frac{1}{2}\frac{1}{s+j\omega_0} - \frac{1}{2}\frac{1}{s-j\omega_0}$$

where $\pm j\omega_0$ are the complex conjugate pair of poles of the system function. Using the first-order tabulated transform pairs we obtain:

$$
\begin{aligned}
y(t) &= u(t) - \frac{1}{2}u(t)e^{-j\omega_0 t} - \frac{1}{2}u(t)e^{j\omega_0 t} \\
&= u(t)\left[1 - \frac{e^{j\omega_0 t} + e^{-j\omega_0 t}}{2}\right] \\
&= u(t)\left[1 - \cos\omega_0 t\right]
\end{aligned}
$$

The simple result in this case has been obtained in, to say the east, a rather long-winded way, particularly when then second-order denominator in $Y(s)$ can be so readily identified with a tabulated second-order transform. We might remedy this by recombining the second and third factors over a common denominator to obtain:

$$Y(s) = \frac{1}{s} - \frac{s}{s^2 + \omega_0^2} \tag{6.20}$$

We may now take advantage of the second-order terms in the table of Laplace transform pairs (Table 6.1) and write the inverse transform as:

$$
\begin{aligned}
y(t) &= u(t) - u(t)\cos\omega_0 t \\
&= u(t)\left[1 - \cos\omega_0 t\right]
\end{aligned}
$$

but there is still considerable redundancy in this procedure. A better approach would be to seek a *partial* partial fraction expansion in terms which allow the use of both first *and* second-order transform pairs (or even higher order as more skill is acquired):

$$Y(s) = \frac{K_1}{s} + \frac{K_2 s + K_3}{s^2 + \omega_0^2}$$

Notice that the numerator for the second-order denominator must be first-order in s. We determine the coefficients K by the method of comparison of coefficients, which appears to be fairly standard in the elementary teaching of this topic. Recombining over a common denominator:

$$Y(s) = \frac{K_1(s^2 + \omega_0^2) + K_2 s^2 + K_3 s}{s(s^2 + \omega_0^2)}$$

and comparing coefficients of s^r:

$$
\begin{aligned}
s^2 &: \quad K_1 + K_2 = 0 \\
s^1 &: \quad K_3 = 0 \\
s^0 &: \quad K_1\omega_0^2 = \omega_0^2
\end{aligned}
$$

Therefore $K_1 = 1$ and $K_2 = -1$, giving the result in equation (6.20) directly and the inverse transform in one further step. The behaviour of the system is shown in Figure 6.4.

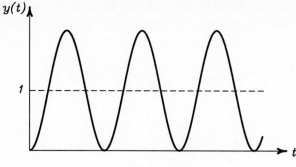

Figure 6.4

It was a straightforward matter in the last example to identify and use an appropriate second-order transform pair. It would not be so obvious how to proceed for a more realistic (finite Q) L CR circuit, but by 'completing the square' in the denominator we may again take advantage of the second-order transform pairs.

Example 6.7 The L CR circuit dealt with in Problems 5.3 and 5.4 has the voltage transfer function:

$$H(s) = \frac{\omega_0^2}{s^2 + s\omega_0/Q + \omega_0^2}$$

where $\omega_0^2 = 1/LC$ and $Q = 1/\omega_0 CR$. When subject to the unit step $u(t)$, the Laplace transform of the output is:

$$Y(s) = \frac{1}{s} \frac{\omega_0^2}{s^2 + s\omega_0/Q + \omega_0^2}$$

By the method of the comparison of coefficients we may factorize this to:

$$Y(s) = \frac{1}{s} - \frac{s + \omega_0/Q}{s^2 + s\omega_0/Q + \omega_0^2}$$

The second-order factor is not immediately amenable to inverse transformation using the tabulated transforms, but by adding and subtracting a suitable term in the denominator we may draft it into such a form:

$$
\begin{aligned}
Y(s) &= \frac{1}{s} - \frac{s + \omega_0/Q}{(s + \omega_0/2Q)^2 + \omega_0(1 - 1/4Q^2)} \\
&= \frac{1}{s} - \frac{s + \omega_0/2Q}{(s + \omega_0/2Q)^2 + \omega_0(1 - 1/4Q^2)} \\
&\quad - \frac{1}{2Q\sqrt{1 - 1/4Q^2}} \frac{\omega_0\sqrt{1 - 1/4Q^2}}{(s + \omega_0/2Q)^2 + \omega_0(1 - 1/4Q^2)}
\end{aligned}
$$

where in the second stage we have separated the terms in the numerator also to be able to identify second-order transform pairs. Using such transforms we obtain the result that:

$$y(t) = u(t) \left[1 - e^{-\gamma t} \cos \omega_m t - \frac{1}{\sqrt{4Q^2 - 1}} e^{-\gamma t} \sin \omega_m t \right]$$

where $\gamma = \dfrac{\omega_0}{2Q}$ and $\omega_m = \omega_0 \sqrt{1 - \dfrac{1}{4Q^2}}$

It may readily be shown that as $Q \rightarrow \infty$ the result for the L C circuit is obtained.

The same result as that in the last example could have been obtained by working with a partial fraction expansion in first-order terms, two of which would have complex roots (for $Q > \frac{1}{2}$). The procedure is lengthy and prone to error.

The above procedures must be modified when a particular pole is repeated, that is, at least part of the denominator may be written as $(s - p)^r$. The partial fraction expansion must then contain terms with denominators in ascending powers of $(s - p)$ but we will not give details here.

6.7   The Impulse Response of Networks

The impulse $\delta(t)$ (or, more properly, *unit* impulse) and the response of a system to it have a special place in network analysis. The unit impulse is another of the so-called singularity functions, one of which, the unit step function $u(t)$, we have encountered already. An infinite series of singularity functions are produced by repeated differentiation or integration of $u(t)$. The unit impulse is derived by a single differentiation of $u(t)$ but, because a discontinuity is involved, we must approach this by means of a mathematical limit. We regard the unit step as a ramp of short duration Δt (Figure 6.5(a)) such that:

$$u(t) = \lim_{\Delta t \to 0} f(t)$$

The differential coefficient of $f(t)$ is as shown in Figure 6.5(b), a pulse of duration Δt and amplitude $1/\Delta t$. It follows that the area of df/dt is unity. The unit area is preserved in the limit:

$$\delta(t) = \lim_{\Delta t \to 0} \frac{df(t)}{dt}$$

leading to the somewhat confusing specification of $\delta(t)$: infinite height, zero width but unit area. That is, we may write for $\delta(t)$:

$$\int_{0-}^{\infty} \delta(t) dt = 1 \tag{6.21}$$

from which it is apparent that the impulse has dimensions of reciprocal time $(\sec^{-1})$.

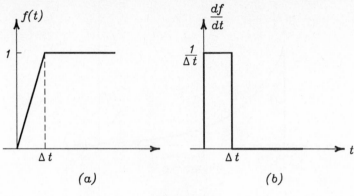

Figure 6.5

For any system with zero initial stored energy and subject to the input $\delta(t)$ we have, taking $F(s)$ from Table 6.1:

$$X(s) = 1 \quad \text{and} \quad Y(s) = H(s)$$

It follows that the particular response for the input $\delta(t)$, usually assigned the symbol $h(t)$, bears a fundamental relationship to the system function:

$$h(t) = \mathcal{L}^{-1}[H(s)] \quad \text{or} \quad \mathcal{L}[h(t)] = H(s)$$

$h(t)$ and $H(s)$ are a Laplace transform pair.

We saw in Chapter 1 that a time domain description of the behaviour of a system can be given in terms of a convolution integral:

$$y(t) = \int_0^\infty x(\tau)h(t-\tau)d\tau$$

which is written symbolically as :

$$y(t) = x(t) \otimes h(t)$$

in which $\otimes$ is the symbol for convolution. We have seen more recently that:

$$Y(s) = X(s)H(s)$$

and now that:

$$\mathcal{L}[h(t)] = H(s)$$

Evidently convolution in the time domain is equivalent to taking the product of Laplace transforms in the s-plane.

Example 6.8 Finding the response of the RC circuit to $\delta(t)$ is very straightforward:

$$H(s) = \frac{Y(s)}{X(s)} = \frac{1/CR}{s+1/CR} \quad \text{with} \quad X(s) = 1$$

therefore $$Y(s) = \frac{1}{CR}\frac{1}{s+1/CR}$$

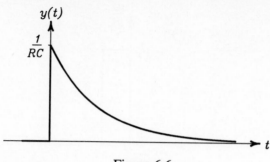

Figure 6.6

The inverse transform is read from Table 6.1:

$$y(t) = u(t)\frac{1}{CR}e^{-\alpha t} \quad \text{with} \quad \alpha = 1/CR$$

and is shown in Figure 6.6. One may readily show that, for the output:

$$\int_{0-}^{\infty} y(t)\,dt = 1$$

The integral of the signal is the same for the output as for the input, as one would expect for a passive circuit.

Example 6.9 For the L C circuit the response is obtained from:

$$H(s) = \frac{X(s)}{Y(s)} = \frac{\omega_0^2}{s^2 + \omega_0^2} \quad \text{with} \quad X(s) = 1$$

so

$$Y(s) = \frac{\omega_0^2}{s^2 + \omega_0^2}$$

for which the inverse transform may be written directly:

$$y(t) = \omega_0 \sin \omega_0 t$$

and is shown in Figure 6.7.

It would be reassuring to find for the output of the L C circuit to the unit impulse that:

$$\int_{0-}^{\infty} y(t)\,dt = 1 \tag{6.22}$$

as it does for the integral of the unit impulse itself, but difficulties would arise in the integration from the infinite limit. Instead the circuit must be seen as a limiting case of the more realistic L CR circuit, dealt with in Example 6.7, for which it would be found that:

$$y(t) = \frac{\omega_0^2}{\omega_m}e^{-\frac{\omega_0}{2Q}t} \sin \omega_m t$$

with the same definitions as for step response. It can again be shown that, for $Q \to \infty$, the response approximates the response of a pure L C circuit. Further, the integration in equation (6.22) would be found to be unity in the limit of $Q \to \infty$.

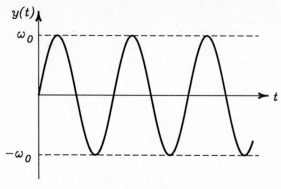

Figure 6.7

6.8 Non-zero Initial Stored Energy

In the case of zero initial stored energy we have seen that the Laplace transformation of the fundamental $i - v$ relationships for the capacitor and the inductor reduce to simple ratios of Laplace transforms, equations (6.12) and (6.14), defining the impedance. When there is stored energy we may retain the impedance as the ratio of the Laplace transforms of voltage and current but augment the model by the addition of extra sources to represent the stored energy (Figure 6.8). The application of Kirchhoff's laws to the models of Figure 6.8 will produce the transform $i - v$ relationships, equations (6.11) and (6.13). Furthermore the two possible models for each component will be seen to be connected by a source transformation (Chapter 1) involving the impedance of the circuit element. One may determine the output for any circuit containing additional sources representing the stored energy in the same way as for any circuit with more than one source, by the application of the principle of superposition. In order to invoke the use of the voltage divider in such calculations one would opt for the voltage source representation of the initial energy. The appropriate diagram for the RC circuit, the capacitor having initial

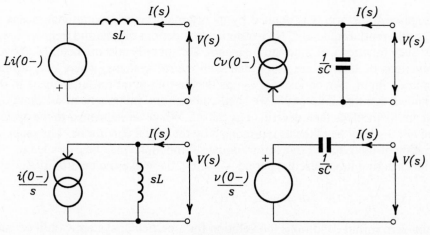

Figure 6.8

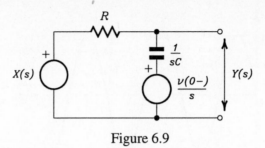

Figure 6.9

stored energy, is therefore as in Figure 6.9. The response of this circuit to an input $x(t)$, the transform of which is $X(s)$, may therefore be obtained by superposing its contribution and that of the initial stored energy source. The contribution of each source is found by the application of a potential divider involving the component impedances defined for zero initial stored energy. That is:

$$
\begin{aligned}
Y(s) &= \frac{1/sC}{R+1/sC}X(s) + \frac{R}{R+1/sC}\frac{v(0-)}{s} \\
&= \frac{1}{sCR+1}X(s) + \frac{CR}{sCR+1}v(0-) \\
&= \frac{\alpha}{s+\alpha}X(s) + \frac{1}{s+\alpha}v(0-) \quad \text{with} \quad \alpha = \frac{1}{CR}
\end{aligned}
\qquad (6.23)
$$

We have already examined the solution with, for example, $v(0-) = 0$ and $x(t) = u(t)$ (the ZSR for the unit step input). When $x(t) = 0$ but $v(0-) \neq 0$, that is the input is shorted after the initial conditions are established, we shall observe the *zero input response* (ZIR) of the system, what we have previously called the free response. It is straightforward to show from equation (6.23) that, with $v(0-) = 1$ but $X(s) = 0$ because $x(t) = 0$ then the ZIR is $u(t)e^{-\alpha t}$, agreeing with the earlier result obtained in Chapter 5.

6.9   The Bilateral Laplace Transform

The Laplace transform was defined by its originator as an operational means of solving differential equations. No restriction to functions of time and frequency was made. Such functions, of time and frequency, are our only concern here. Moreover, the functions of interest are associated with *causal* systems, where output does not precede input. We choose $t = 0-$ as the starting point for an analysis in the sure knowledge that for $t \geq 0-$ we shall embrace the full extent of all functions which are themselves then described as causal. Whatever happened to the system before $t = 0-$ may be taken as represented by the initial conditions. The range of integration of the Laplace transform for causal functions may be extended to $-\infty$ without changing it because the period $-\infty < t < 0-$ makes no contribution:

$$
F(s) = \int_{0-}^{\infty} f(t)e^{-st}dt \quad \rightarrow \quad \int_{-\infty}^{\infty} f(t)e^{-st}dt
\qquad (6.24)
$$

For non-zero initial conditions the solution for a particular system would be augmented, as described in section 6.8, by the addition of sources equivalent to the period $t < 0-$. The form of the transform with the symmetrical range of integration is said to be the *bilateral Laplace transform*. As a means of distinguishing the two transforms the original form is often called the *unilateral Laplace transform*. We shall later see a close connection between the bilateral Laplace transform and the Fourier transform, indeed the latter is a special case of the former.

Summary

In this chapter we examined a wider use of the exponential factor e^{st} (where s is the complex frequency) than its application simply to solve the homogeneous differential equations which describe the free response of systems. It is now seen as the *kernel* of the *Laplace transform* which converts differential and integral equations to an algebraic form so facilitating a general solution. The use of the Laplace transform in this way is described as an *operational calculus*, so-called because it seeks to deal with the *operations* of calculus by alternative means. Another such method was originated by Oliver Heaviside but is not now widely used. Operational methods are by no means confined to calculus; indeed the humble logarithm is an example of an operational method in arithmetic calculation.

The Laplace transforms of the operations of integration, differentiation and so on were examined, as were the transforms of common functions: the unit step, the unit impulse and damped and steady sinusoids. Proceeding from the differential equations for simple networks, the ratio of the Laplace transform of the output to the Laplace transform of the input was deduced. *For the case of zero initial stored energy* that ratio could be interpreted as the system function $H(s)$ obtained previously in the more limited context of the exponential input e^{st}.

Taking the Laplace transform of the simple differential equations for the current–voltage characteristics of a capacitor and an inductor we obtained the ratio of the Laplace transform of voltage to the Laplace transform of current, again in the case of zero initial stored energy, of $1/sC$ for the capacitor and sL for the inductor. Such factors had been described as the *impedances* when dealing with complex exponential inputs. We now give them a wider interpretation as the ratio of Laplace transforms and the description *generalized* or *Laplacian impedances*. The generalized impedance of a resistor is simply its resistance. Again, applying the concept of the impedance potential divider to treating networks comprising capacitors, inductors and resistors allows the system function $H(s)$ to be obtained without reference to the system differential equation.

To find the time domain output for a network subject to a particular input requires that, having obtained the Laplace transform of the output as the product of the system function, usually found by reference to the component impedances, and the Laplace transform of the input, inverse transformation must be performed. In many cases the required inverse transform may be deduced by *partial fraction expansion* of the output transform into simple factors followed by reference to tabulated transforms. *Long division* of numerator by denominator may sometimes be necessary to render the output transform in a form where partial fraction expansion is possible, and sometimes a method of *completing the square* in second-order denominators may save the labour of reducing the transform to first-order terms. When an inverse transform cannot be found by reference to tables of transform pairs it can, in principle, be evaluated by performing an integration in the complex s-plane, but this is beyond the scope of an elementary treatment.

While the case of zero initial stored energy may be dealt with simply by using component impedances, an allowance for non-zero initial stored energy in components can be made in an equally straightforward way by representing any particular component which has initial energy using an ideal source in conjunction with the usual impedance. The value of each ideal source is determined by the initial conditions as they relate to a particular component. Source transformation allows us to choose either a voltage source or a current source representation.

Problems

6.1 Show that the Laplace transform of the unit step function is $1/s$, that is, show:

$$\int_{0-}^{\infty} u(t)e^{-st}dt = 1/s$$

6.2 Find the zero state response (ZSR) of the circuit in Figure 6.10 when the voltage source is the unit step function.

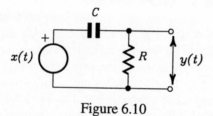

Figure 6.10

6.3 Find the zero state response of the network in Figure 6.11 when the voltage source is the unit step function.

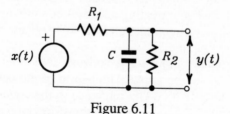

Figure 6.11

6.4 Use the method of Laplace transforms to find the zero state response of the circuit in Figure 6.12 to the input $u(t)\cos\omega t$.

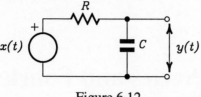

Figure 6.12

6.5 Show, by 'completing the square' in the denominator of the Laplace transform of the output, that the ZSR of the L CR circuit shown in Figure 6.13 to an input which is the unit impulse is:

$$y(t) = \frac{\omega_0^2}{\omega_m} e^{-\frac{\omega_0}{2Q}t}\sin\omega_m t \quad \text{where} \quad \omega_m = \omega_0\sqrt{1 - \frac{1}{4Q^2}}$$

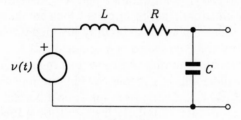

Figure 6.13

7

The Fourier Series and Fourier Transform

7.1 Time and Frequency Domains

We referred in Chapter 4 to the preoccupation in electronic engineering with sine waves even though they carry no information and will therefore not be processed individually by the information-handling circuits which are the major object of our study. The reason for such emphasis can be fully justified through an understanding of Fourier's theorem which will soon be explained, but at least as important as theoretical justification is our intuition. We know that in an orchestra the violin (or possibly the piccolo) will be able to generate the highest note of all. This is top G, a sine wave of about 3200 Hz, arising from the simple harmonic motion (s.h.m.) of a string (or a column of air). Similarly there will be a lowest note, from the double bass, lower C, about 32 Hz. We know intuitively that if we construct a circuit which performs satisfactorily at these two frequencies and at every frequency between then it will deal satisfactorily with the orchestra as a whole. Intuition suggests that it would be more difficult to take the signal in its entirety, a function of time, and calculate directly how it is changed by the circuit to a new function of time. From the practical point of view of circuit synthesis, it would generally be necessary to carry such a calculation to very long times to ensure that the required time response was obtained, whereas the behaviour with respect to the various *frequency components* remains true at all times (for a time-invariant system). That is, it appears intuitively easier to engineer the *frequency response* of a circuit than to engineer the time response. The two possible approaches are distinguished by saying that we work in either the *time domain* or the *frequency domain*. The specification of a signal in one domain prescribes the description of the signal in the other. The relationship between the two domains is shown schematically in Figure 7.1. The fact that we have the option of working in either domain is a great concession.

Fourier's theorem relies for its application on the property of sine functions that they are transmitted by a linear system without change. In mathematical language one says that sine functions are eigen (or characteristic) functions of a linear system. Finally, the superposition principle guarantees that the components specified by Fourier can in fact be added. We see that both the eigen property and the

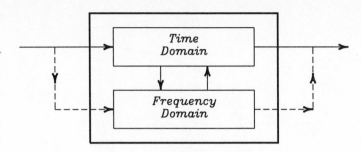

Figure 7.1

superposition of functions follow from the linearity of circuits.

In summary, therefore, we may say that Fourier specifies which functions should be added while linearity determines that they can be added.

7.2 Fourier's Theorem

Towards the end of the eighteenth century and at the beginning of the nineteenth, a prime concern of mathematicians was the representation of an arbitrary function by a series of analytic functions, such as a series of sines and cosines. Much work had already been done on the description of the displacement of a stretched string by Bernoulli, d'Alembert and Euler. Fourier somewhat shattered the air of steady application by announcing, in 1807, that *any* function defined in a fixed interval could be replaced by an infinite sum of harmonic sine and cosine functions. He could not prove his assertion, and indeed it was another 22 years before Dirichlet detailed the exact conditions under which Fourier's statement could be made. However contentious it may have been when first given, Fourier's theorem has provided one of the cornerstones of modern mathematics and is one of the dominant themes in network analysis.

Fourier himself was concerned to analyze functions connected with heatflow problems and so confined his attention to a function in a particular interval. It is usual now to specify the interval as being from $-\pi$ to $+\pi$ (Figure 7.2). The sum of harmonic sine and cosine functions, the Fourier series, is:

$$f(x) = \frac{a_0}{2} + \sum_{n=1}^{n=\infty} (a_n \cos nx + b_n \sin nx) \tag{7.1}$$

The factor of $\frac{1}{2}$ in the zero-order coefficient means that the same general expression obtains as for the higher-order coefficients which are given (for proof see Appendix A) as follows:

$$a_n = \frac{1}{\pi} \int_{-\pi}^{\pi} f(x) \cos nx \, dx \tag{7.2}$$

$$b_n = \frac{1}{\pi} \int_{-\pi}^{\pi} f(x) \sin nx \, dx \tag{7.3}$$

As the lowest harmonic is periodic in the interval 2π it follows that the series not only reproduces the function in the interval $-\pi$ to $+\pi$ but in every succeeding

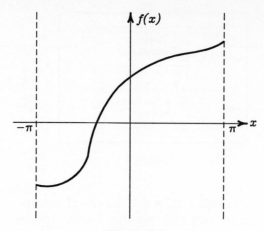

Figure 7.2

interval of 2π as well. It is said to produce the *periodic extension* of the function. In the present discussion we are very much concerned with periodic functions and so the periodic extension coincides with our needs. Indeed we shall adopt the usual convention of considering the theorem as being stated for periodic functions. If the function that we wish to analyze is not periodic in any interval, then we could select a particular interval for the application of the Fourier series. We would have to ignore the function outside that range and accept that the Fourier series produces the periodic extension of the chosen range. Alternatively, we could employ a development of the basic theorem, the Fourier integral, which allows the range to be extended to infinity. There is no dispute that this idea is wholly attributable to Fourier and it is presented later.

Economy in the calculation of coefficients is possible if there is some symmetry in the function. For instance, that shown in Figure 7.3 (as a periodic extension) is an *odd* function, $f(x) = -f(-x)$, and will therefore require only *odd* functions (sines) in its Fourier sum.

Example 7.1 Let us calculate the zero-order coefficient and the first coefficient for the harmonic series of sines for the function shown in Figure 7.3. We perform the

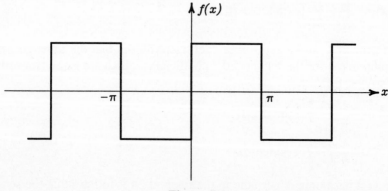

Figure 7.3

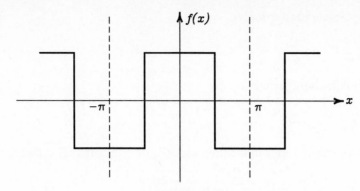

Figure 7.4

integrals in two segments, from $-\pi$ to 0 and from 0 to π:

$$a_0 = \frac{1}{\pi}\left\{\int_{-\pi}^{0}(-1)\,dx + \int_{0}^{\pi}(1)\,dx\right\} = 0$$

$a_0/2$ is the average level (the d.c. level) of the function and it is self-evident that this should be zero.

$$\begin{aligned} b_1 &= \frac{1}{\pi}\left\{\int_{-\pi}^{0}(-1)\sin x\,dx + \int_{0}^{\pi}(1)\sin x\,dx\right\} \\ &= \frac{1}{\pi}\left\{[\cos x]_{-\pi}^{0} + [-\cos x]_{0}^{\pi}\right\} = \frac{4}{\pi} \end{aligned}$$

Similarly, further coefficients are found to be:

$$b_3 = \frac{4}{3\pi}, \quad b_5 = \frac{4}{5\pi}, \quad b_7 = \frac{4}{7\pi}, \text{ etc.}$$

Notice for this function that, while we need only the b coefficients, even then we require just those of odd order.

With a different choice of axes (a shift of $\pi/2$, Figure 7.4) we could regard the above function as *even*, $f(x) = f(-x)$, and it would then be given as a series of even functions (cosines). Recognizing that $\cos(x - \pi/2) = \sin x$ it should come as no surprise that this is so.

We have drafted the theorem above in terms of the spatial variable x to accord with the historical development, but from the point of view of electronic signals *time* is the appropriate variable. Furthermore, the interval of interest for a periodic function is usually given as the period T rather than 2π. We will therefore redraft the theorem with a simple change of variable obtained by scaling the interval 2π to T. First we write the Fourier series for a function of the time variable t' in the interval $-\pi \leq t' \leq \pi$ (and periodic in 2π):

$$g(t') = \frac{a_0}{2} + \sum_{n=1}^{n=\infty}(a_n\cos nt' + b_n\sin nt')$$

We now use simple proportion:

$$\frac{t'}{t} = \frac{2\pi}{T}$$

to produce a substitution for t':

$$t' = \frac{2\pi}{T}t$$

enabling us to write a series in the time variable t for the interval $-T/2 \leq t \leq T/2$:

$$f(t) = \frac{a_0}{2} + \sum_{n=1}^{n=\infty} \left(a_n \cos n\frac{2\pi}{T}t + b_n \sin n\frac{2\pi}{T}t\right)$$

(It is pedantic but correct to use two different symbols f and g for the two functions since the function $g(t')$ does *not* evaluate correctly if values of t are used and vice versa for $f(t)$, unless $T = 2\pi$, which is trivial.) Now $1/T$ is the frequency (ν_0) of the lowest frequency component (the fundamental) and so the corresponding angular frequency ω_0 is given by:

$$\omega_0 = 2\pi\nu_0 = \frac{2\pi}{T}$$

Substitution in the above produces the following form of the Fourier series:

$$f(t) = \frac{a_0}{2} + \sum_{n=1}^{n=\infty} \left(a_n \cos n\omega_0 t + b_n \sin n\omega_0 t\right)$$

with the coefficients now given as:

$$a_n = \frac{2}{T} \int_{-T/2}^{T/2} f(t) \cos n\omega_0 t \, dt$$

$$b_n = \frac{2}{T} \int_{-T/2}^{T/2} f(t) \sin n\omega_0 t \, dt$$

In general, therefore, as illustrated in Figure 7.5, a signal, a function in the *time domain* will produce *two* sets of numbers in the *frequency domain*, the coefficients a_n and b_n. They are the amplitudes of the sine and cosine components of the signal and may be described as its spectra.

A more compact and useful form of the Fourier series may be obtained by making use of the Euler expression to replace the sine and cosine functions with complex exponentials thus:

$$
\begin{aligned}
f(t) &= \frac{a_0}{2} + \sum_{n=1}^{n=\infty} \left[a_n \frac{e^{jn\omega_0 t} + e^{-jn\omega_0 t}}{2} + b_n \frac{e^{jn\omega_0 t} - e^{-jn\omega_0 t}}{2j} \right] \\
&= \frac{a_0}{2} + \sum_{n=1}^{n=\infty} \left[\frac{a_n - jb_n}{2} e^{jn\omega_0 t} + \frac{a_n + jb_n}{2} e^{-jn\omega_0 t} \right] \\
&= c_0 + \sum_{n=1}^{n=\infty} \left[c_n e^{jn\omega_0 t} + c_{-n} e^{j(-n)\omega_0 t} \right]
\end{aligned}
$$

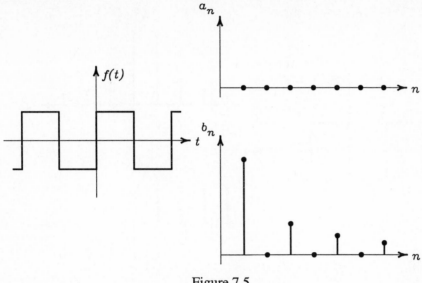

Figure 7.5

where we have introduced the zero-order ($n = 0$) term c_0 and the complex coefficients $\mathbf{c}_n$ and $\mathbf{c}_{-n}$, all defined in terms of a_n and b_n thus:

$$\mathbf{c}_{\pm n} = \frac{a_n \mp jb_n}{2} \tag{7.4}$$

Furthermore, rather than perform the sum from 1 to ∞ for pairs of terms, one with positive n the other with negative n, it is equivalent to perform the sum for one term from $-\infty$ to ∞:

$$f(t) = \sum_{n=-\infty}^{n=\infty} \mathbf{c}_n e^{jn\omega_0 t}$$

The coefficients $\mathbf{c}_n$ may be shown to be given as follows:

$$\mathbf{c}_n = \frac{1}{T} \int_{-T/2}^{T/2} f(t) \exp - jn\omega_0 t \, dt$$

and can be given in the Euler form:

$$\mathbf{c}_n = c_n e^{j\Phi_n}$$

They are the *phasors* of the various components of the signal. We saw earlier that the trigonometric form of the Fourier series produces two spectra in the frequency domain for a signal in the time domain, the amplitudes a_n and b_n of the cosine and sine terms. The complex exponential form of the Fourier series also yields two spectra in the frequency domain but these are now a set of amplitudes and a set of phases, the moduli and arguments of the complex coefficients $\mathbf{c}_n$. The spectra for a square wave are illustrated in Figure 7.6 where a certain symmetry will be noted. The symmetry follows from the requirement that an electronic signal be represented

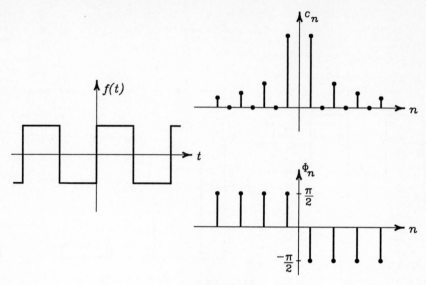

Figure 7.6

by a wholly real function. Every component of the trigonometric Fourier series must therefore be wholly real and so the coefficients of the exponential form must occur in complex conjugate pairs (equation (5.2)). Furthermore, if the function is even then the need for only cosine terms in the trigonometric sum ($b_n = 0$) reduces equation (7.4) to:

$$c_{\pm n} = \frac{a_n}{2}$$

The coefficients are wholly real and are shown on a phasor diagram (for an arbitrary value of n) in Figure 7.7(a). Similarly, for an odd function the coefficients must be imaginary:

$$c_{\pm n} = \mp j\frac{b_n}{2}$$

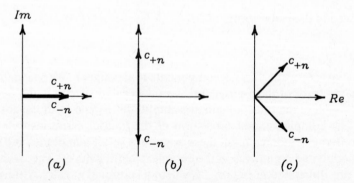

Figure 7.7

and are shown in Figure 7.7(b). The case for a function which is neither even nor odd (both cosine and sine terms in the trigonometric series) is shown in Figure 7.7(c).

7.3 The Fourier Transform

It is quite often the case that we have or wish to discuss a phenomenon which occurs but once in an infinite time-span, for instance a pulse or a step function. In this case we certainly do not require a representation which attempts a periodic extension of the function. We produce the required representation from the complex form of the Fourier series by first introducing the explicit form of the coefficients into the complex Fourier sum:

$$f(t) = \sum_{n=-\infty}^{n=\infty} \left[\frac{1}{T} \int_{-T/2}^{T/2} f(t') \exp-jn\omega_0 t' \, dt' \right] \exp jn\omega_0 t \qquad (7.5)$$

We use t' as a dummy variable within the inner integral to make it clear over which variable the integral is taken. Now we allow the range $-T/2$ to $T/2$ to increase to $-\infty$ to ∞ so preserving the symmetry in the time interval with respect to $t = 0$. As a consequence the interval between frequency components (ω_0) becomes infinitesimally small:

$$\lim_{T \to \infty} \left[\omega_0 = \frac{2\pi}{T} \right] = d\omega$$

Using this to substitute for $1/T$ in equation (7.5) and at the same time writing the frequency of the nth component ($n\omega_0$) simply as ω we obtain:

$$f(t) = \int_{-\infty}^{\infty} \frac{d\omega}{2\pi} \left[\int_{-\infty}^{\infty} f(t') \exp-j\omega t' \, dt' \right] \exp j\omega t$$

where the sum from $-\infty$ to ∞ for infinitesimally small increments has been replaced by the integral. The inner integral is a function of frequency and we give this the symbol $F(j\omega)$. We therefore have (dropping the prime on t):

$$f(t) = \frac{1}{2\pi} \int_{-\infty}^{\infty} F(j\omega) \exp j\omega t \, d\omega \qquad (7.6)$$

$$\text{with} \quad F(j\omega) = \int_{-\infty}^{\infty} f(t) \exp-j\omega t \, dt \qquad (7.7)$$

The latter is the Fourier transform, the former the inverse Fourier transform. They link functions, one continuous in time, the other continuous in frequency. The Fourier transform, $F(j\omega)$, is the counterpart of the coefficients c_n in the exponential form of the Fourier series except that the multiplier $1/T$ has been reincorporated in $f(t)$ to provide the incremental step in the limit of infinite steps.

It is appropriate at this stage to make clear the close connection between the Laplace and Fourier transforms which was mentioned earlier. We may do so by considering those particular complex frequencies s for which the real part is zero, so that $s = j\omega$. We are then concerned with the complete range of constant amplitude

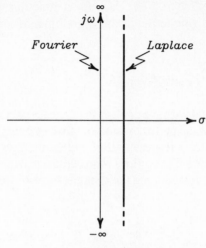

Figure 7.8

sinusoids (steady a.c.), and the bilateral form of the Laplace transform (equation (6.24)) becomes:

$$F(j\omega) = \int_{-\infty}^{\infty} f(t)e^{-j\omega t}dt$$

which is the Fourier transform just defined. Similarly, for steady a.c. the inverse Laplace transform becomes:

$$f(t) = \frac{1}{2\pi} \int_{-\infty}^{\infty} F(j\omega)e^{j\omega t}d\omega$$

which is the inverse Fourier transform. The inverse Fourier transform may therefore be described as an inverse Laplace transform performed on the imaginary axis in the s-plane (Figure 7.8), whereas the inverse Laplace transform is performed along a line parallel to the imaginary axis suitably placed in the right-hand s-plane. As intimated in Chapter 6, values of s in the right-hand s-plane ensure that the Laplace transform involves an exponential decay and can therefore always be made to converge. No more can be said here about convergence other than that the ability to ensure convergence is an important advantage that the Laplace transform has with respect to the Fourier transform. There are many functions for which no Fourier transform can be found but for which a Laplace transform exists.

A shorthand, as for the Laplace transform, is usually adopted and we represent the transform and its inverse symbolically using $\mathcal{F}$ and $\mathcal{F}^{-1}$, so that:

$$\mathcal{F}[f(t)] = F(j\omega) \quad \text{and} \quad \mathcal{F}^{-1}[F(j\omega)] = f(t)$$

Example 7.2 Let us find, as an example, the Fourier transform of the causal exponential $u(t)(1/RC)\exp-(t/RC)$. In the use here of the term 'causal' we wish to attribute to the function properties such as we would expect to be associated with the output of a causal system, excitation of which commences at $t = 0$ and before

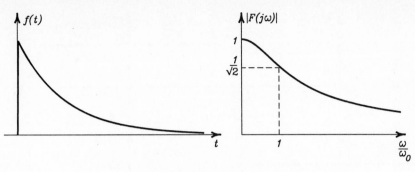

Figure 7.9

which time we would expect no output. A causal function is therefore one which has zero value before $t = 0$ and this is signified by the multiplier $u(t)$. The property is crucial to being able to evaluate the integral in the Fourier transform:

$$F(j\omega) = \int_{-\infty}^{\infty} u(t)\omega_0 \exp-\omega_0 t \exp-j\omega t \, dt \quad \text{with} \quad \omega_0 = \frac{1}{RC}$$

Since $u(t)$ is zero before $t = 0$ we may adjust the limits of integration and write:

$$
\begin{aligned}
F(j\omega) &= \omega_0 \int_0^{\infty} \exp-(\omega_0+j\omega)t \, dt \\
&= \frac{\omega_0}{-(\omega_0+j\omega)}[\exp-(\omega_0+j\omega)t]_0^{\infty} \\
&= \frac{\omega_0}{-(\omega_0+j\omega)}(0-1) \\
&= \frac{\omega_0}{\omega_0+j\omega} \quad \text{or} \quad \frac{1}{1+j\omega/\omega_0}
\end{aligned}
$$

The modulus of the transform is:

$$|F(j\omega)| = \frac{1}{\sqrt{1+\omega^2/\omega_0^2}}$$

which is shown, together with the causal function, in Figure 7.9. They will later be recognized respectively as the frequency response of a first-order low-pass filter and the impulse response of such a filter.

7.4 Impulses in Time and Frequency: Duality and Convolution

Having now established the close connection between the Fourier and Laplace transforms it should come as no surprise that the unit impulse and the unit impulse response, so central to our discussion of the Laplace transform in regard to convolution and the system function, have an equally important postion in relation to the Fourier transform. Because functions of a frequency which is steady a.c. are

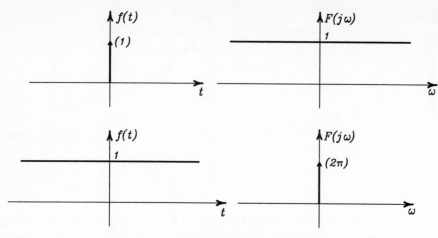

Figure 7.10

easier to grasp than functions of a complex frequency, direct practical significance can be given to many of the emerging ideas and these are examined in detail in Chapter 8.

Let us start by finding the Fourier transform of the unit impulse $\delta(t)$ which, by definition, occurs at $t = 0$:

$$X(j\omega) = \int_{-\infty}^{\infty} \delta(t)e^{-j\omega t}\, dt$$

$\delta(t)$ is only non-zero at $t = 0$ when $\exp - j\omega t$ is unity. Therefore:

$$X(j\omega) = \int_{-\infty}^{\infty} \delta(t)\, dt = 1 \qquad (7.8)$$

by reference to the definition of $\delta(t)$ (Chapter 6).

It is equally pertinent to consider the function $\delta(\omega)$, an impulse in the frequency domain. It is the 'frequency spectrum' of a signal having a constant amplitude at zero frequency and no other, that is, d.c. In fact we consider $2\pi\delta(\omega)$ and find the associated time-domain behaviour:

$$x(t) = \frac{1}{2\pi} \int_{-\infty}^{\infty} 2\pi\delta(\omega)e^{-j\omega t}\, d\omega$$

$\delta(\omega)$ is only non-zero at $\omega = 0$ when $\exp - j\omega t = 1$. So:

$$x(t) = \int_{-\infty}^{\infty} \delta(\omega)\, d\omega = 1$$

again by reference to the definition of the unit impulse (the change of variable from t to ω in no way affects the definition).

The two results we have just obtained are displayed in Figure 7.10 where an obvious symmetry will be seen. We may capitalize on this symmetry by once again noting that the precise style of the variables is irrelevant to the definition of the Fourier transform and so transpose ω and t in the transform (and its inverse):

$$F(jt) = \int_{-\infty}^{\infty} f(\omega)\exp - j\omega t\, d\omega \tag{7.9a}$$

$$f(\omega) = \frac{1}{2\pi}\int_{-\infty}^{\infty} F(jt)\exp j\omega t\, dt \tag{7.9b}$$

which transforms a function $F(jt)$ of t to a function $f(\omega)$ of ω, at the same time defining the inverse. If we now reverse the sign of either variable (let us choose ω first) and rearrange factors of 2π, equation (7.9b) becomes:

$$2\pi f(-\omega) = \int_{-\infty}^{\infty} F(jt)\exp - j\omega t\, dt$$

Alternatively, reversing the sign of t in equation (7.9a) and dividing both sides by 2π:

$$\frac{1}{2\pi}F(-jt) = \frac{1}{2\pi}\int_{-\infty}^{\infty} f(\omega)\exp j\omega t\, d\omega$$

The first of the derived equations is a Fourier transform, the second is an inverse transform. We therefore see that if $F(j\omega)$ is the Fourier transform of $f(t)$ then $2\pi f(-\omega)$ is the transform of $F(jt)$ or, equivalently, $(1/2\pi)F(-jt)$ is the inverse transform of $f(\omega)$. Both statements express the time/frequency *duality* of the Fourier transform.

Example 7.3 When $f(t) = \delta(t)$ we found earlier that the Fourier transform is 1. Now rather than find the time-domain behaviour for an impulse in frequency by performing an *inverse* transformation (which would be simple enough in this case) we can apply the duality principle and infer that for a function which is unity for all positive and negative time the Fourier transform is $2\pi\delta(\omega)$. The same result was obtained earlier (by inverse transformation).

An important application of the unit impulse (Dirac delta function) is in the definition of the operation of convolution of functions in the time domain (Chapter 1):

$$y(t) = \int_{-\infty}^{\infty} x(\tau)h(t-\tau)d\tau$$

which is written symbolically as:

$$y(t) = x(t) \otimes h(t)$$

where $\otimes$ is the symbol for convolution. We saw when dealing with the Laplace transform that the behaviour in the complex frequency (s) plane is described by a *product* of Laplace transforms. Not surprisingly, an analogous result is obtained for the Fourier transform which we may show explicitly by taking the transform throughout of the convolution integral:

$$\mathcal{F}[y(t)] = Y(j\omega) = \int_{-\infty}^{\infty}\left[\int_{-\infty}^{\infty} x(\tau)h(t-\tau)\, d\tau\right]e^{-j\omega t}\, dt$$

Multiplying and dividing by $e^{j\omega\tau}$ within the integral:

$$Y(j\omega) = \int_{-\infty}^{\infty} \int_{-\infty}^{\infty} x(\tau)h(t-\tau)e^{-j\omega(t-\tau)}e^{-j\omega\tau} \, d\tau \, dt$$

we may now regroup terms to produce the product of two integrals:

$$Y(j\omega) = \int_{-\infty}^{\infty} x(\tau)e^{-j\omega\tau} \, d\tau \int_{-\infty}^{\infty} h(t-\tau)e^{-j\omega(t-\tau)} \, dt$$

In the first we integrate over τ and the result is a function of ω; in the second we integrate over t, for a particular τ, and the result is also a function of ω. To make the result clearer we could change the origin of t by τ in the second integral, using instead the variable $t' = t - \tau$ with $dt' = dt$:

$$Y(j\omega) = \int_{-\infty}^{\infty} x(\tau)e^{-j\omega\tau} \, d\tau \int_{-\infty}^{\infty} h(t')e^{-j\omega t'} \, dt'$$

Remarking yet once again that the variable over which the integration is taken is irrelevant to the definition of the Fourier transform, what we have here is the product of two transforms:

$$Y(j\omega) = X(j\omega) \, H(j\omega) \tag{7.10}$$

where $X(j\omega) = \mathcal{F}[x(t)]$ and $H(j\omega) = \mathcal{F}[h(t)]$. Convolution in the time domain therefore requires the product of Fourier transforms in the frequency domain. The transform of $h(t)$, $H(j\omega)$, describes the amount by which we must scale every frequency component in the input, $X(j\omega)$, to obtain every component in the output; it is the *frequency response* of the system and the steady a.c. counterpart of $H(s)$. Most importantly $H(j\omega)$ is the Fourier transform of the impulse response of the system and we see again that the impulse response gives us complete information about a system.

For purposes of analysis we may sometimes wish to discuss the convolution of frequency response functions in the frequency domain. Recalling what has already been said about defining the Fourier transform *from* the frequency domain *to* the time domain, it could be shown by following the above argument but interchanging time and frequency, that convolution of frequency response functions is equivalent to taking the product of the associated responses in the time domain.

7.5 Time-shifted Impulses and Impulse Sequences

We now extend our discussion to include an impulse, $\delta(t - t_0)$, occurring at $t = t_0$ rather than just at $t = 0$ (Figure 7.11(a)). Its Fourier transform is:

$$X(j\omega) = \int_{-\infty}^{\infty} \delta(t-t_0)e^{-j\omega t} \, dt$$

Multiplying and dividing by $e^{j\omega t_0}$ within the integral:

$$X(j\omega) = \int_{-\infty}^{\infty} \delta(t-t_0)e^{-j\omega(t-t_0)}e^{-j\omega t_0} \, dt$$

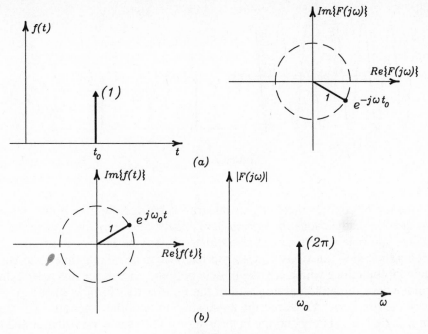

Figure 7.11

Shifting the origin of t by t_0 and using again the variable $t' = t - t_0$ with $dt' = dt$:

$$X(j\omega) = e^{-j\omega t_0} \int_{-\infty}^{\infty} \delta(t') e^{-j\omega t'} \, dt'$$

The factor $e^{-j\omega t_0}$ is not a function of t' and so precedes the integral which is the Fourier transform of $\delta(t)$, that is, unity. Therefore:

$$X(j\omega) = e^{-j\omega t_0}$$

Taking advantage of time/frequency duality we can immediately write for an impulse in the frequency domain, $2\pi\delta(\omega - \omega_0)$, that the associated time domain function is $e^{j\omega_0 t}$ (Figure 7.11(b)).

Example 7.4 Combining two delta functions in the frequency domain shifted equal amounts in positive and negative frequency:

$$X(j\omega) = \pi\delta(\omega - \omega_0) + \pi\delta(\omega + \omega_0)$$

we use the previous result, itself obtained by resort to the duality principle, to find:

$$
\begin{aligned}
x(t) &= \frac{1}{2}(e^{j\omega_0 t} + e^{-j\omega_0 t}) \\
&= \cos\omega_0 t
\end{aligned}
$$

The result, and its transform, are shown in Figure 7.12. We may note that a periodic function needs *two* delta functions in the time domain for its description. The direct

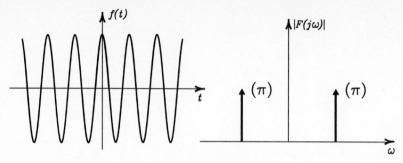

Figure 7.12

transformation of $\cos \omega_0 t$, involving an integration between infinite limits, would require esoteric mathematics to perform but is made simple by the application of the duality principle. The example also illustrates that the Fourier transform, far from being limited to functions which occur but once in an infinite time-span, may be applied to functions which are indefinitely periodic. It was for this reason that we avoided describing the development of the Fourier transform as intended just for aperiodic functions. However, the application to periodic functions may then require a degree of subtlety and the Fourier series will usually be better suited to the problem.

A series of delta functions with time, or frequency, shifts in arithmetic progression is described as a *comb function*. The comb function has great practical significance in the sampling of continuous time signals. The Fourier transform of a time domain comb function is very interesting and may be obtained by considering first a finite duration delta sequence extending from $-NT$ to NT where T is the interval between impulses (Figure 7.13(a)). For each impulse the Fourier transform is an exponential decay, so that:

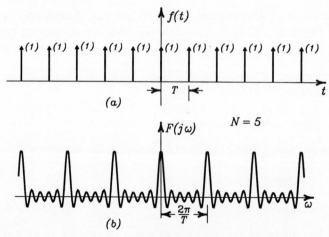

Figure 7.13

$$\mathcal{F}\left[\sum_{n=-N}^{n=N}\delta(t-nt_0)\right] = F(j\omega) = \sum_{n=-N}^{n=N}e^{-jn\omega t}$$

The standard approach to summing a geometric series produces:

$$F(j\omega) = \frac{e^{jN\omega t}-e^{-j(N+1)\omega t}}{1-e^{-j\omega t}}$$

Multiplying both numerator and denominator by $e^{j\omega t/2}$ we go on to find:

$$F(j\omega) = \frac{e^{j(N+\frac{1}{2})\omega t}-e^{-j(N+\frac{1}{2})\omega t}}{e^{j(\omega t)/2}-e^{-j(\omega t)/2}}$$

$$= \frac{\sin\left(N+\frac{1}{2}\right)\omega t}{\sin(\omega t)/2}$$

Each time ωt approaches 2π we may use the approximation $\sin\theta \approx \theta$ in both the numerator and the denominator:

$$F(j\omega) \rightarrow \frac{\left(N+\frac{1}{2}\right)\omega t}{\omega t/2} = 2N+1$$

The Fourier transform is therefore a series of peaks of height $(2N+1)$ occurring at frequency intervals of $2\pi/T$ (Figure 7.13(b)). The width of each major peak is $2\pi/(N+\frac{1}{2})T$ so that in the limit of an infinite comb function the major peaks will approach infinite height but zero width, and these are the characteristics of a delta sequence. The Fourier transform of an infinite comb function in one domain is therefore an infinite comb function in the other domain. We shall make use of this result in a major example when we have examined one more type of function.

7.6   Rectangular Pulses in Time and Frequency

Pulses of finite duration, not delta functions, are frequently employed in signalling and their use in radar is well known. The Fourier transform of a rectangular pulse in time defined in the interval $-T/2 > t > T/2$ (Figure 7.14(a)) is:

$$F(j\omega) = \int_{-T/2}^{T/2}(1)e^{-j\omega t}\,dt$$

$$= \frac{1}{-j\omega}(e^{-j\omega t/2}-e^{j\omega t/2})$$

$$= T\left(\frac{\sin\omega t/2}{\omega t/2}\right) \qquad (7.11)$$

The bracketed terms of the form $(\sin x)/x$ are usually described as a *sinc* function.

As a rectangular function symmetrically disposed in the time domain corresponds to a sinc function in the frequency domain, time/frequency duality allows us to say that a sinc function in the time domain will correspond to a symmetrical

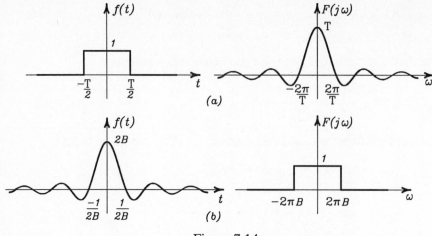

Figure 7.14

rectangular function in the frequency domain (Figure 7.14(b)). Such a function describes for positive frequencies the frequency response of a filter which passes low frequencies but not high frequencies with a sharp cut-off between the two (an ideal low-pass filter of bandwidth B).

Example 7.5 To illustrate the use of some of the preceding ideas consider a system which band-limits a signal (using an ideal low-pass filter) to a bandwidth B Hz in the positive frequency domain, a total bandwidth of $2B$ Hz, samples it by multiplication with a delta sequence where the pulse interval is $1/(2B)$ sec^{-1} and then seeks to recover the information in the original signal by the use of a further ideal low-pass filter (Figure 7.15(a)). Multiplication of a signal by a function in the time domain is, we know, equivalent to convolution of the Fourier transform of the signal (its frequency spectrum, Figure 7.15(b)) with the Fourier transform of the multiplying function. We also know that the Fourier transform of an infinite delta sequence in time is itself an infinite delta sequence in frequency. The effect of convolving a single delta function with the frequency response is to leave the response function unchanged because as we scan the delta function through the spectrum it selects each value in turn:

$$\int_{-\infty}^{\infty} H(j\omega_0)\delta(\omega - \omega_0)\, d\omega_0 = H(j\omega)$$

Convolution with a delta sequence where the frequency interval is $1/(2B)$ replicates the frequency spectrum indefinitely (Figure 7.15(c)). The effect of the recovery filter is to truncate the frequency spectrum once again to $\pm B$, restoring the original spectrum (Figure 7.15(d)). It may seem that this is not a very surprising result but it is in fact remarkable. We have shown that if a band-limited signal is sampled at an appropriate rate then we may recover the signal *with exactly the same frequency spectrum* despite the fact that we have temporarily dispensed with much of the information in the signal and taken only samples of it. If the recovered signal has exactly the same spectrum as the original signal we may reasonably say that it is

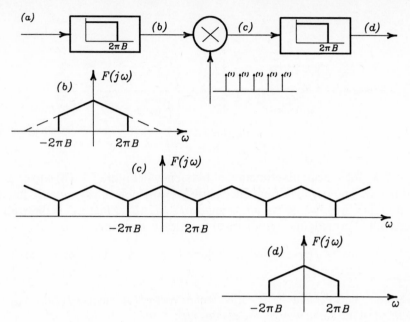

Figure 7.15

the same signal. We have given an adequate proof of what is known as the sampling theorem which, in its various forms, bears the names of Nyquist, Bode and Shannon. A proof based on the time-domain behaviour is laborious. It describes the impulse response of the recovery filter (a sinc function) as 'interpolating' between the various signal samples; the sum of the impulse responses to the signal samples will reconstruct the original signal. We are here of course treading the interface between continuous time and discrete time (or sample data) signal processing, the latter also known as digital signal processing. From whichever side the interface is approached, a complete understanding of sampling and reconstruction will only be obtained from an understanding of the (continuous time) Fourier transform.

Finally in regard to rectangular pulses we may produce the transform of functions shifted in time, as we did for the impulse function, and extend this to the frequency domain by resort to time/frequency duality, again as we did for the impulse. We found that the transform of a time-shifted impulse is:

$$\mathcal{F}[\delta(t - t_0)] = e^{-j\omega_0 t}$$

The result may readily be shown to apply to all functions. That is:

$$\mathcal{F}[f(t - t_0)] = e^{-j\omega_0 t} F(j\omega) \tag{7.12}$$

Calling on time/frequency duality we find:

$$\mathcal{F}[2\pi f\{-(\omega + \omega_0)\}] = e^{-j\omega_0 t} F(jt) \tag{7.13}$$

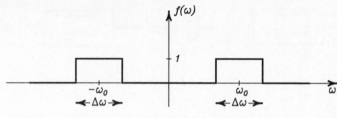

Figure 7.16

Example 7.6 The rectangular function of frequency in Figure 7.14(b) when subject to equal postive and negative frequency shifts of ω_0 yields the symmetric function shown in Figure 7.16, where we are using $\Delta\omega/2 = 2\pi B$ so that $\Delta\omega$ is the width of each rectangle. Formally the symmetrical function is the sum:

$$f(-\omega - \omega_0) + f((-\omega + \omega_0)), \quad \text{where} \quad \begin{aligned} f(-\omega) &= 1; \quad |\omega| < \Delta\omega/2 \\ &= 0; \quad |\omega| > \Delta\omega/2 \end{aligned}$$

The function of ω is even and it is equivalent to write $f(\omega - \omega_0) + f((\omega + \omega_0))$ with $f(\omega)$ defined in the same way as for $f(-\omega)$.

Combining the result of the application of time/frequency duality to time-shifted functions, equation (7.13), with the Fourier transform of a rectangular pulse in time, equation (7.11), we find the transform of the pair of rectangular pulses in frequency:

$$
\begin{aligned}
\mathcal{F}[f(\omega - \omega_0) + f((\omega + \omega_0))] &= \frac{1}{2\pi}\left[\Delta\omega\frac{\sin\frac{1}{2}\Delta\omega t}{\frac{1}{2}\Delta\omega t}e^{-j\omega_0 t} + \Delta\omega\frac{\sin\frac{1}{2}\Delta\omega t}{\frac{1}{2}\Delta\omega t}e^{j\omega_0 t}\right] \\
&= \frac{\Delta\omega}{\pi}\frac{\sin\frac{1}{2}\Delta\omega t}{\frac{1}{2}\Delta\omega t}\cos\omega_0 t \qquad (7.14)
\end{aligned}
$$

The result is a *modulated* sinusoid. It is a non-causal signal in that it exists in both positive and negative time. It is not practically realizable but then neither is the frequency-domain function from which it is derived. The latter is the frequency response of an ideal band-pass filter. However, the exercise still allows the observation that a modulated signal will require *two* identical frequency bands in its frequency spectrum. These are the *side-bands* defined in the discussion of modulation. Here we can see them, as for the case of a sum of time-shifted delta functions, to be the amplitudes of the conjugate pairs of complex exponentials which together describe the (wholly real) function of time.

Summary

While the complex frequency dealt with in previous chapters may allow detailed *analysis* of circuits, it only makes sense in *practical* situations to discuss periodic voltages and currents which are steady in amplitude, *steady a.c.* In fact we surmise that if we can arrange for a system to have a satisfactory response to each and every frequency component in a signal then the response of the system to the signal as a whole will be equally satisfactory. In this we are making an intuitive application of the principle of superposition. We see working in the *frequency domain* as distinct from working in the *time domain*.

A major mathematical theorem derived by *Fourier* completely supports our intuition. It shows how to represent a function defined in the interval $-\pi$ to π in terms of a series of sine and cosine functions. The periodic nature of sine and cosine naturally produces a *periodic extension* of the function which exactly coincides with our need to be able to deal with periodic, but non-simple harmonic, functions and shows us how to decompose them into a series of sine and cosine functions. Two sets of information in the frequency domain are available for any signal in the time domain, the *amplitudes* of the sine and cosine components.

A mathematically more useful and sophisticated form of Fourier's theorem expresses a function as a sum of complex exponential factors, complex numbers in their Euler form. Each factor then has the role of a phasor for a component of the signal. Again, two series of data in the frequency domain are available regarding the time domain signal, the *amplitude* and *phase* of each phasor. The exponential form of the Fourier series provides a very useful point from which to develop a statement of Fourier's theorem for a function, defined in the interval $-\infty$ to ∞, in terms of an infinite set of frequency components at infinitesimally small frequency intervals, the *Fourier integral* or *Fourier transform*. The Fourier transform allows us to deal with functions which are defined on an infinite time-scale and are generally not periodic like those dealt with using the Fourier theorem. One particular non-periodic function, the *unit impulse*, is seen to be of special importance in that the frequency components of which it is comprised extend over the complete frequency spectrum. The response of a system to the unit impulse, the *impulse response*, therefore carries much information about a system.

Recognizing the fact that the precise choice of symbol is unimportant in the fundamental definition of a mathematical expression, we transposed ω and t in the definition of the Fourier transform to produce its *dual*. In this way one can determine without further work the transform for other functions which are dual to the first. For instance, the transform of a totally steady function in time, d.c., is seen to be an impulse at $\omega = 0$ in the frequency domain, the dual of the impulse in time described earlier. How time-shifts of functions behave during transformation is also relevant particularly for the impulse. An infinite series of impulses with time shifts in arithmetic progression, a so-called *comb function*, has a Fourier transform which, somewhat surprisingly, is also a comb function. Multiplying a function by a comb function in the time domain is a formal way of describing the *sampling* of a function. A function in the frequency domain comprising two bands of frequencies shifted equal amounts in positive and negative frequency with respect to zero was seen to correspond to a *modulated* signal. The symmetrical function of frequency describes the *side-bands* of the modulated signal or conversely the response of an ideal band-pass filter. The impulse response of the filter would be the modulated function seen in the time domain.

The Fourier transform also leads to a useful alternative view of what is difficult to visualize in the time domain, the convolution of two functions. Usually the functions convolved are the input and the impulse response, for the purpose of predicting the output. Convolution in the time domain is seen to translate into the *product* of Fourier transforms in the frequency domain so producing the Fourier transform of the output. The Fourier transform of the input is its frequency spectrum, likewise for the output, making clear that the Fourier transform of the impulse response is that function which translates the signal spectrum from input to output, the *frequency response* of the system. The duality of the transform also allows us to say that if convolution in the time-domain translates into a product in the frequency domain then convolution in the frequency domain will translate into a product in the time domain.

Combining our knowledge of sampling using comb functions, the transformation of convolution and duality, one can account in a compact way for how a sample data system with appropriate input and output filters can produce an exact replica of the original signal despite the fact that only samples of the signal are processed.

Problems

7.1 Sketch the frequency spectrum of a continuous sinusoid of constant amplitude (a 'pure' sine function) the frequency of which is 10 Hz. Reasoning intuitively, how would the spectrum differ from the pure sinusoid case if, instead, the sine function were lightly damped?

7.2 Calculate the amplitude of the lowest frequency a.c. component of the full-wave rectified sine function shown in Figure 7.17(a). What is the lowest frequency component of the half-wave rectified sine function, shown on an identical time-scale, in Figure 7.17(b)? (No further calculation is necessary.)

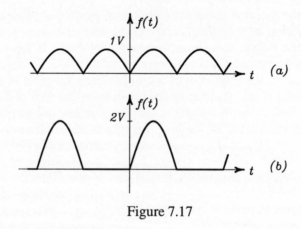

Figure 7.17

7.3 Examine the sawtooth waveform shown in Figure 7.18 and explain whether only sine terms or only cosine terms or both will be required in its Fourier expansion. (No calculation is necessary.) State, without proof, the value of the zero-order Fourier coefficient and explain its significance. Calculate the amplitude of the lowest frequency a.c. component.

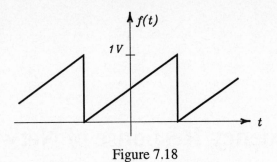

Figure 7.18

7.4 Find the Fourier transform of the symmetrical triangular function shown in Figure 7.19.

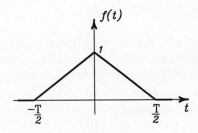

Figure 7.19

7.5 Prove the result in equation (7.12).

7.6 Obtain the result given in equation (7.14) by performing an inverse Fourier transform on the function defined in Figure 7.16.

8

The Frequency Response of Networks

8.1   The Response to Steady a.c.

The most important result of our foray into operational methods (principally the Laplace transform) has been the development of a method, involving so-called device impedances, for finding the system function without reference to the system differential equation. Circuits could be solved for certain archetypal inputs, for instance $u(t)$ and $\delta(t)$, in a completely general way. In principle, the same method could be extended to the input 'Hello' where the circuit might be, say, a pre-amplifier. In fact it would be impracticable, and indeed virtually impossible, to carry out such an analysis, not to say highly unnecessary. The analysis would produce no guarantee that the circuit would work satisfactorily for the signal 'Goodbye', for which an entirely new analysis would be required. While we should (and must) have the general methods of analysis at our disposal, we in fact seldom require the complete analysis of a circuit for a particular input. Our immediate objective is usually to obtain rather more cursory information about the performance of a circuit. Does the circuit boost low frequencies relative to high frequencies? Will the pre-amplifier treat all frequencies equally up to a certain maximum and not change their relative phases? It is at this level that we operate as we produce and evaluate *designs* for new circuits. That is, armed with knowledge from our analysis about the properties of circuits we now proceed to the first stages of *synthesis*, the course of action suggested in Chapter 1. What we require for the first step in design is the general specification of a system.

The specification of a circuit, required as the precursor to implementation, is given by stating the desired response of the circuit to a range of frequencies of steady a.c. Quoting the specification in this way is entirely in line with our intuitive reaction supported by Fourier's theorem as detailed in Chapter 7, to express any signal in terms of a set of steady a.c. components. More importantly, it is the only practicable way to specify the circuit given that test instruments, which could be used to verify the design, typically generate and display steady a.c. components. Stated more formally, we are proposing, for purposes of specification, to restrict the more general Laplace transform to those particular values of s for which $\sigma = 0$ (hence *steady* a.c.), that is, we are proposing to use a Fourier transform. We can then give intuitively satisfying descriptions to both the transform of the signal, $F(j\omega)$ – it is the amplitude and phase spectrum of the signal – and to the system function, $H(j\omega)$ – it is the frequency response of the system. By the latter we mean the way that the system responds to each steady a.c. component to which it is subject. The way in which the frequency spectrum $F(j\omega)$ relates to the frequency response $H(j\omega)$ determines the performance of a circuit in a particular situation.

In regard to the proposed use of the Fourier transform, it should go without saying that it is never possible to satisfy the required limits of integration: that is, we can never wait an infinite time. For a symphony the signal may last an hour, for spacecraft communication many years, but not an infinite time. It is now that we apply some common sense. Outside of our interval of interest (the hour of the symphony) it is not unreasonable to say that the signal is zero to ∞ and was zero from $-\infty$ before the signal commenced. In this way we could, in principle, calculate the Fourier transform to find the frequency spectrum. An equally valid approach would be to use the Fourier series (suitable for application to a limited time interval) to obtain a sum over a number of steady a.c. components. We might interpolate between components to produce a continuous frequency spectrum as for the Fourier transform. Now more common sense: we know without detailed calculation what, in general terms, either approach will yield. For a symphony orchestra this would be a range of frequencies from a few hertz to about 20 kHz (above that the characteristic of the human ear is a limitation anyway). For Radio 1 on FM we would have a band of frequencies 200 kHz wide centred on 98.8 MHz. What we propose, therefore, is to work in the spirit rather than the detail of Fourier's theorem. It is often the case that once one has examined the details of a piece of theory it becomes clear how to use it in broad outline.

Example 8.1 The type of general description of a circuit which we require are shown in Figure 8.1. The response shown in Figure 8.1(a) would pass all the frequencies of the symphony orchestra without change (a good pre-amplifier), while the response of Figure 8.1(b) would isolate Radio 1 from other stations and noise (a good tuner).

We have taken a little while here to place steady a.c. analysis in the context of the more general analysis introduced earlier. The matter is quite often dealt with rather informally. From what has already been said it follows that a major part of

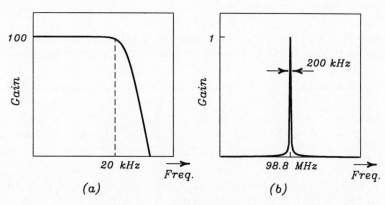

Figure 8.1

this chapter will be devoted to specifying networks in terms of their frequency response. It will in fact be sufficient for our purpose to confine our attention to first- and second-order networks. We shall write the response function for the network with the minimum of components in each case and that then serves as an archetype for the response. The use of any such network would indicate an intention to alter the frequency spectrum of a signal and so the network merits the description 'filter'. Indeed, any network, perhaps not intended as a filter, will change the frequency spectrum between input and output. The archetypal responses will serve to model such networks. Responses of higher order can always be obtained by cascading low-order modules. Before examining these low-order filter responses we consolidate the idea of impedance in the case of steady a.c. As in the case of the generalized impedance we shall then be able to write down response functions almost by inspection using principally the concept of the impedance potential divider. It may come as an agreeable surprise to the reader to find that the material with which we have now to deal is simpler than that in recent chapters.

8.2 The Steady a.c. Impedance

We can, of course, obtain the frequency response $H(j\omega)$ by writing $s = j\omega$ in the system function $H(s)$ appropriate to a particular circuit. Bearing in mind that $H(s)$ could itself be obtained by the application of simple network equations, usually just the potential divider, involving the complex impedance $Z(s)$, we might expect to define steady a.c. impedances $Z(j\omega)$ and proceed to $H(j\omega)$ directly. That is, for the resistor we still write R but for the inductor sL becomes $j\omega L$ and for the capacitor $1/sC$ becomes $1/j\omega C$ or, in rational form, $-j/\omega C$. These impedances figure so widely in the subject that it is worthwhile confirming their derivation independently. Following closely the procedure in Chapter 5, we apply a voltage source (Figure 8.2), $v(t) = \mathbf{v} \exp j\omega t$, and find the current developed, $i(t) = \mathbf{i} \exp j\omega t$. The exponential now takes as the coefficient of t in the argument, the wholly imaginary complex frequency $j\omega$, representing steady a.c. ($\mathbf{i}$ and $\mathbf{v}$ are the current and voltage phasors). It is implicitly understood that only the real part of each complex source is required. For a resistor (Figure 8.2(a)):

$$\mathbf{v}e^{j\omega t} = R\mathbf{i}e^{j\omega t}$$

so that $\quad \mathbf{v} = R\mathbf{i}$ \hfill (8.1)

The resistance R is wholly real and indicated by a point on the real axis of an Argand diagram (Figure 8.3(a)). The current and voltage phasors therefore differ

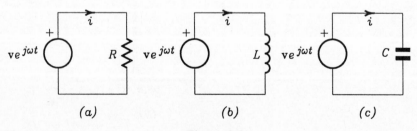

(a) (b) (c)

Figure 8.2

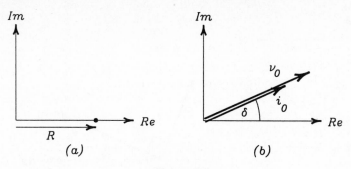

Figure 8.3

only in amplitude. On a phasor diagram (Figure 8.3(b)) they would both occur at the same angular position corresponding to the initial phase.

For an inductor (Figure 8.2(b)):

$$\mathbf{v}e^{j\omega t} = L\frac{d}{dt}\mathbf{i}e^{j\omega t}$$
$$= j\omega L\mathbf{i}e^{j\omega t}$$

so that $\mathbf{v} = j\omega L\mathbf{i}$ (8.2)

The ratio of phasors, $j\omega L$, is, in this case, wholly imaginary and so represented by a point on the positive imaginary axis of an Argand diagram (Figure 8.4(a)). The magnitude of this ratio, ωL, corresponds to the ratio of the amplitudes of the current and voltage phasors. In regard to the factor j it may be recalled from the discussion of the Argand diagram in Chapter 5 that multiplication by j predicates a rotation of one phasor with respect to another of 90°. It may be seen directly that such a phase is introduced by writing j in the Euler form:

$$j = e^{j\pi/2}$$

Expansion in the usual way will show that this is so. Equation (8.2) may then be written:

$$\mathbf{v} = \omega L\mathbf{i}e^{j\pi/2}$$

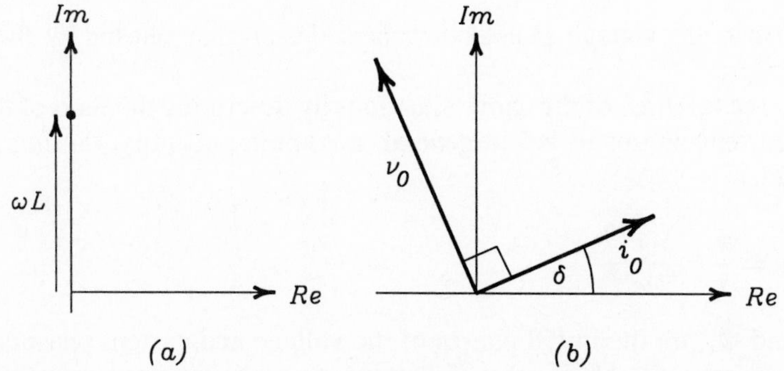

Figure 8.4

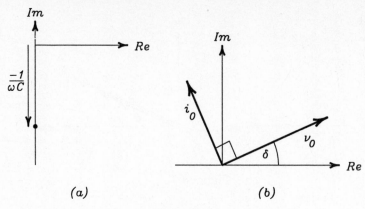

(a)   (b)

Figure 8.5

and it is clear that the voltage phasor *leads* the current phasor by 90° (Figure 8.4(b)).
Lastly, for a capacitor (Figure 8.2(c)):

$$\mathbf{i}e^{j\omega t} = C\frac{d}{dt}\mathbf{v}e^{j\omega t}$$
$$= j\omega C\mathbf{v}e^{j\omega t}$$

so that $\mathbf{v} = \dfrac{1}{j\omega C}\mathbf{i}$ (8.3)

or in rational form

$$\mathbf{v} = -j\frac{1}{\omega C}\mathbf{i}$$ (8.4)

Once again the ratio of phasors, $-j1/(\omega C)$, is wholly imaginary but, in this case,
represented by a point on the negative imaginary axis of an Argand diagram (Figure 8.5(a)). The magnitude of the ratio, $1/(\omega C)$, corresponds, as for the inductor,
to the ratio of the amplitudes of the current and voltage phasors. Making use of the
Euler form for $-j$ we may redraft equation (8.3) as follows:

$$\mathbf{v} = \frac{1}{\omega C}\mathbf{i}e^{-j\pi/2}$$

and so see that the voltage phasor now lags the current phasor by 90° (Figure 8.5(b)).

We embrace all three of the above situations by describing the ratio of the voltage to the current phasor to be, in general, a complex quantity, the *impedance*,
$Z(j\omega)$. That is:

$$Z(j\omega) = \frac{\mathbf{v}}{\mathbf{i}} = \frac{v_0 e^{j\Phi_2}}{i_0 e^{j\Phi_1}}$$

where Φ_2 and Φ_1 are the initial phases of the voltage and current phasors (as introduced in Chapter 5). Notice that, following a widely accepted convention, we
use bold symbols only for those complex quantities which are phasors. For other

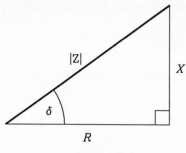

Figure 8.6

complex quantities it will be necessary to distinguish the modulus in the usual way. Expressing the impedance in its Euler form:

$$Z(j\omega) = |Z|e^{j\delta} \tag{8.5}$$

and bringing the two previous equations together:

$$|Z|e^{j\delta} = \frac{v_0 e^{j\Phi_2}}{i_0 e^{j\Phi_1}} = \frac{v_0}{i_0}e^{j(\Phi_2 - \Phi_1)} \tag{8.6}$$

we see that the modulus of $Z(j\omega)$ is the ratio of amplitudes of voltage to current, (v_0/i_0), while the argument of $Z(j\omega)$, δ, is the phase difference $(\Phi_2 - \Phi_1)$ between the voltage and current.

8.3   Driving-point Impedance and Admittance

We drew attention in Chapter 4 to the relevance in their respective areas of application of the one-port and the two-port. For the two-port we focus attention on the transfer function. For the single pair of terminals of the one-port we define a driving-point function which will be a driving-point impedance or admittance depending upon whether we require the voltage to current ratio or vice versa. The driving-point functions are sometimes also called the *input* impedance or *input* admittance. When the context is clear, as in what follows, we may simply say the impedance or admittance.

It is, of course, entirely equivalent to the Euler form, equation (8.5), to express the impedance in its cartesian form:

$$Z(j\omega) = R + jX \tag{8.7}$$

In the case of a resistance, the impedance is just that resistance, hence the choice of symbol for the real part. An inductor or a capacitor presents only *reactance*. For an inductor, the inductive reactance (X_L) is ωL, while for a capacitor the capacitative reactance (X_C) is $-1/\omega C$. On an Argand diagram, the resistive and reactive parts are two sides of what is often called the *impedance triangle* (Figure 8.6). The hypotenuse is the magnitude of the impedance. From the impedance triangle it may be seen that the Euler and cartesian forms are related as follows:

$$|Z| = \sqrt{R^2 + X^2}, \quad \tan\delta = \frac{X}{R}$$

The inverse of the impedance is the *admittance*, $Y(j\omega)$:

$$Y(j\omega) = \frac{\mathbf{i}}{\mathbf{v}}$$

$$= \frac{1}{R + jX}$$

$$= G + jB$$

where G is the *conductance* and B is the *susceptance*. Rationalizing the reciprocal of impedance (multiplying top and bottom by the complex conjugate $R - jX$) we obtain:

$$Y(j\omega) = \frac{R}{R^2 + X^2} + j\frac{-X}{R^2 + X^2}$$

$$= \frac{R}{|Z|^2} - j\frac{X}{|Z|^2} \tag{8.8}$$

Evidently:

$$G = \frac{R}{|Z|^2} \quad \text{and} \quad B = -\frac{X}{|Z|^2}$$

Similarly:

$$R = \frac{G}{|Y|^2} \quad \text{and} \quad X = -\frac{B}{|Y|^2}$$

where $|Y|^2 = G^2 + B^2$. It is important to appreciate that, while the impedance and admittance may be the reciprocal one of the other, the real and imaginary parts are by no means so simply related. Our algebra is now that of *complex quantities* and both real and imaginary parts of the impedance contribute to the real part of the admittance and also to the imaginary part of the admittance. Only in the case of pure resistance, when $X = 0$, may we write $G = R^{-1}$.

Example 8.2 In Figure 8.7 we show a resistor with series capacitor to which is applied a voltage source. KVL requires:

$$v_R(t) + v_C(t) - \mathbf{v}e^{j\omega t} = 0$$

Expressing $v_R(t)$ and $v_C(t)$ separately in terms of the component impedances and the current phasor we have:

$$R\mathbf{i}e^{j\omega t} + jX_C\mathbf{i}e^{j\omega t} - \mathbf{v}e^{j\omega t} = 0$$

We may rearrange the above to obtain the ratio of phasors (the impedance):

$$Z = \frac{\mathbf{v}}{\mathbf{i}} = R + jX_C$$

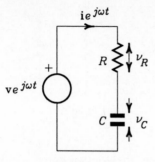

Figure 8.7

using $X_C = -1/(\omega C)$

$$Z = R - j\frac{1}{\omega C} \tag{8.9}$$

demonstrating that we sum impedances in series as we do resistances. In terms of the modulus and argument of the Euler form, equation (8.5):

$$|Z| = \sqrt{R^2 + (1/\omega C)^2}, \quad \tan\delta = \frac{-1}{\omega CR} \tag{8.10}$$

Both of these quantities are frequency-dependent (Figure 8.8). The modulus is dominated at low frequency by the high impedance of the capacitance and at high frequencies, where the impedance of the capacitance is low, by the resistance. For similar reasons the voltage phasor lags the current phasor at low frequencies but the phase difference decreases at high frequencies. The phase difference is 45° ($\tan\delta = -1$) when $\omega = 1/CR$. We recognize this as the reciprocal of what we have previously (Chapter 4) called the *time constant* of an RC combination.

For the case of $R = 1000\ \Omega$ in series with a capacitor $C = 1\ \mu F$ at a supply angular frequency of 1000 rad sec^{-1} (approximately 160 Hz):

$$\frac{1}{\omega C} = 1000\ \Omega \quad \text{and} \quad Z = 1000 - j1000\ \Omega$$

At frequencies above 160 Hz the combination will be more resistive, while below 160 Hz it will appear more capacitative.

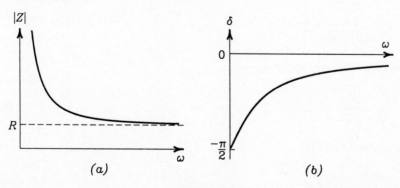

Figure 8.8

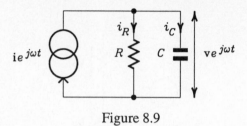

Figure 8.9

Example 8.3 We now consider the application of a current source to the parallel combination of a resistor and capacitor (Figure 8.9). We obtain using KCL:

$$i_R(t) + i_C(t) - \mathbf{i}\,e^{j\omega t} = 0$$

Again using the individual component relationships:

$$\frac{\mathbf{v}e^{j\omega t}}{R} + \frac{\mathbf{v}e^{j\omega t}}{jX_C} - \mathbf{i}e^{j\omega t} = 0$$

So that the admittance:

$$
\begin{aligned}
Y &= \frac{1}{Z} \\
&= \frac{\mathbf{i}}{\mathbf{v}} = \frac{1}{R} + \frac{1}{jX_C}
\end{aligned}
\tag{8.11}
$$

Notice that in finding the admittance of components in parallel we combine the resistive and reactive contributions reciprocally, in a way analogous to the procedure for resistors. In that case the contribution of each resistor to the admittance was pure conductance. Now that we are dealing with resistive *and* reactive components in parallel we sum the wholly real impedances reciprocally (for the conductance) and sum the wholly imaginary impedances reciprocally (for the susceptance). Using $X_C = -1/(\omega C)$ and then rationalizing the second term in equation (8.11):

$$
\begin{aligned}
\frac{1}{Z} &= \frac{1}{R} + j\omega C \\
&= \frac{1 + j\omega CR}{R}
\end{aligned}
$$

The impedance is therefore:

$$Z = \frac{R}{1 + j\omega CR} \tag{8.12}$$

which in rational form is:

$$Z = \frac{R(1 - j\omega CR)}{1 + (\omega CR)^2} \tag{8.13}$$

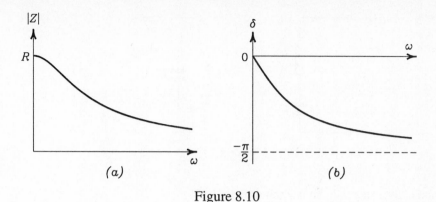

Figure 8.10

It follows that the real (resistive) part of Z is $R/(1+(\omega CR)^2)$ while the imaginary (reactive) part is $-\omega CR^2/(1+(\omega CR)^2)$ and we see that capacitance makes a contribution to the 'resistive' part of the impedance while resistance contributes to the 'reactive' part. In this case the modulus and argument of the Euler form are:

$$|Z| = \frac{R}{\sqrt{1+(\omega CR)^2}}, \qquad \tan\delta = -\omega CR \qquad (8.14)$$

The frequency dependence (Figure 8.10) is now dominated by the lower impedance: the resistance at low frequency and the capacitance at high frequency.

Again taking the case of a 1000 Ω resistor and a capacitor $C = 1\ \mu\text{F}$ at a supply angular frequency of 1000 rad sec^{-1} but now with the components in parallel, the admittance of the resistive component:

$$G = \frac{1}{R} = 10^{-3}\ \text{siemens}$$

while the admittance of the reactive component:

$$jB = \frac{1}{1/j\omega C} = j10^{-3}\ \text{siemens}$$

so that $Y = 10^{-3} + j10^{-3}$ siemens

from which we would find:

$$Z = \frac{10^{-3}}{10^{-6}+10^{-6}} - j\frac{10^{-3}}{10^{-6}+10^{-6}} = 500 + j500\ \Omega$$

In this case at frequencies above 160 Hz the admittance of the combination will be dominated by the capacitance, as will the impedance, while below 160 Hz it will appear more resistive.

Example 8.4 As further examples we now examine series and parallel combinations of an inductor and a capacitor (Figure 8.11) which have interesting features we shall call upon later. Dispensing with applications of KVL and KCL, the impedance

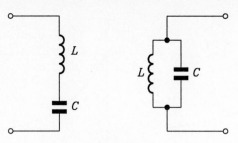

Figure 8.11

of L and C in series may be written directly as:

$$Z = j(X_L + X_C)$$
$$= j(\omega L - 1/\omega C) \tag{8.15}$$

Hence $|Z| = (\omega L - 1/\omega C)$

The impedance falls to zero when $\omega L = 1/\omega C$, that is, when $\omega = 1/\sqrt{(LC)}$ (Figure 8.12(a)). When dealing previously (Chapter 4) with the excitation of circuits we saw this as the frequency of free oscillations. As the real part of the impedance is zero the argument of Z may only be written as a limit:

$$\tan\delta = \lim_{R\to 0} \frac{\omega L - 1/\omega C}{R}$$

and gives the behaviour shown in Figure 8.12(b). For the parallel LC circuit the admittance:

$$Y = \frac{1}{jX_L} + \frac{1}{jX_C}$$

and the impedance:

$$Z = \frac{jX_L\, jX_C}{jX_L + jX_C}$$
$$= \frac{j\omega L/j\omega C}{j\omega L + 1/j\omega C} = \frac{L/C}{j(\omega L - 1/\omega C)} \tag{8.16}$$

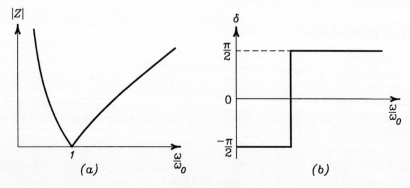

Figure 8.12

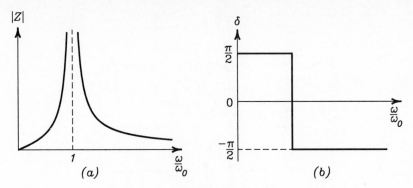

Figure 8.13

The denominator disappears at $\omega = 1/\sqrt{(LC)}$ so the impedance then becomes infinite (Figure 8.13). The behaviour of δ is similarly the inverse of that for the series LC combination.

8.4   Reflected Impedance in Networks with Transformers

The components we have previously identified, in Chapter 4, as transformers are most commonly employed with steady a.c. excitation. Then the situation is usually that a source is applied to one winding while a load is connected to the other (Figure 8.14). For such a steady a.c. source expressed in the usual complex exponential notation, an application of KVL to the primary circuit and to the secondary circuit produces, after dividing throughout by $\exp j\omega t$:

$$\mathbf{v}_1 = j\omega L_1 \mathbf{i}_1 + j\omega M \mathbf{i}_2 \tag{8.17a}$$
$$0 = j\omega M \mathbf{i}_1 + j\omega L_2 \mathbf{i}_2 + Z_2 \mathbf{i}_2 \tag{8.17b}$$

Eliminating $\mathbf{i}_1$ between these two equations, we find $\mathbf{i}_2$ in terms of $\mathbf{v}_1$ and hence:

$$\frac{\mathbf{v}_2}{\mathbf{v}_1} = \frac{-Z_2 \mathbf{i}_2}{\mathbf{v}_1} = \frac{j\omega M Z_2}{\omega^2 (M^2 - L_1 L_2) + j\omega L_1 Z_2} \tag{8.18}$$

Equation (8.17b) yields directly:

$$\frac{\mathbf{i}_2}{\mathbf{i}_1} = \frac{-j\omega M}{j\omega L_2 + Z_2} \tag{8.19}$$

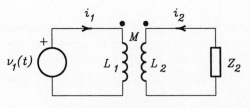

Figure 8.14

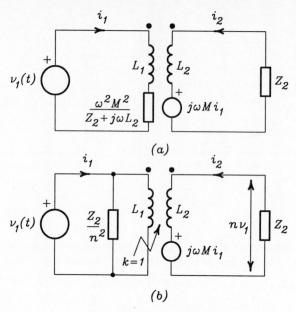

(a)

(b)

Figure 8.15

Example 8.5 A transformer intended for power applications will almost always be subject to the idealization, introduced earlier, of unity coupling. It ensures that $M/L_1 = L_2/M = n$ (the turns ratio) and $M^2 = L_1 L_2$ (Chapter 4). From equation (8.18) therefore we find:

$$\frac{\mathbf{v}_2}{\mathbf{v}_1} = \frac{M}{L_1} = n \tag{8.20}$$

An accompanying condition in such applications is that, by virtue of the amount of core material, the primary and secondary inductances will be large. At the intended frequency of operation the impedance of the primary and of the secondary will be large compared with the load impedance, with the result that:

$$\frac{\mathbf{i}_2}{\mathbf{i}_1} = \frac{1}{n} \tag{8.21}$$

The effect of the load impedance on the primary circuit can be seen by eliminating $\mathbf{i}_2$ between equations (8.17a) and (8.17b) to obtain:

$$\frac{\mathbf{v}_1}{\mathbf{i}_1} = j\omega L_1 + \frac{\omega^2 M^2}{j\omega L_2 + Z_2} \tag{8.22}$$

The impedance of the primary circuit is therefore supplemented by an amount $\omega^2 M^2 / (j\omega L_2 + Z_2)$, *the reflected impedance*, in series with the impedance of the primary inductance (Figure 8.15(a)). It may be represented as a shunt impedance by first combining the terms in equation (8.22) over a common denominator whence:

$$\frac{\mathbf{v}_1}{\mathbf{i}_1} = \frac{j\omega L_1 Z_2}{j\omega L_2 + Z_2}$$

Again employing the result that for unity coupling $L_2/L_1 = n^2$ one finds:

$$\frac{\mathbf{v}_1}{\mathbf{i}_1} = \frac{j\omega L_1 Z_2}{j\omega n^2 L_1 + Z_2} = \frac{j\omega L_1 Z_2/n^2}{j\omega L_1 + Z_2/n^2}$$

The effect is the same as having an impedance Z_2/n^2 in parallel with the primary impedance (Figure 8.15(b)). In the case of the ideal transformer the primary impedance will be very large and so the effective impedance seen by the source will be simply Z_2/n^2. It follows that by an appropriate choice of turns ratio a transformer can effect impedance matching between two otherwise disparate circuits.

The idealizations in the above treatment have been expressed in terms of the flux linkage and the relative size of the winding impedances. Losses in the windings, so-called copper losses, and losses in the core have been taken to be negligible. In a more complete treatment such losses would be accounted for by additional resistance in both primary and secondary circuits.

8.5   Logarithmic Scales, Decibels and Bode Plots

In discussing driving-point impedances we have examined their variation with frequency. The reader may see it as common sense that we shall not require the same detail in an interval between, say, 10 kHz and 100 kHz as we should in the interval between 1 Hz and 10 Hz. Indeed, we will usually specify the same number of points in the two intervals. It is a natural *practical* requirement to work in frequency *decades* (powers of ten) and, accordingly, common instruments (oscilloscopes, signal generators, etc.) are so calibrated. Equal intervals for each frequency decade are produced on a graph by using the logarithm to base ten of the frequency. The same consideration applies to the quantity being plotted. There is equal interest in the ten-fold change from 1 to 10 as there is from 100 to 1000. A logarithmic scale is again appropriate. As an example of the use of such a log-log plot the impedance of the parallel RC combination is shown in Figure 8.16. The frequency response of each of the simple networks to be considered shortly might be given in

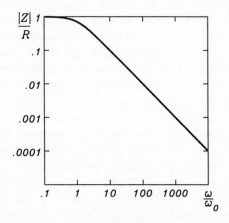

Figure 8.16

terms of any of the ratios listed in Chapter 4: the dimensionless voltage transfer or current transfer ratios, or the transfer impedance or transfer admittance. We shall confine our attention to the first of the dimensionless ratios, the voltage transfer function. However, of even greater practical significance than a voltage ratio is the *power* ratio. Identifying R_o as the resistor in which the output power is developed and R_i similarly for the input, the power ratio may be written:

$$\frac{P_o}{P_i} = \frac{V_o^2/R_o}{V_i^2/R_i}$$

P_o and P_i are the powers, V_o and V_i the r.m.s. voltages, *o* for output, *i* for input. For the case that the powers are developed in equal resistors:

$$\frac{P_o}{P_i} = \frac{V_o^2}{V_i^2} = \left(\frac{V_o}{V_i}\right)^2$$

Adopting a logarithmic scale for the power ratio as for the voltage ratio:

$$\log_{10}\left(\frac{P_o}{P_i}\right) = 2\log_{10}\left(\frac{V_o}{V_i}\right)$$

The unit of this logarithmic ratio is the *bel*. The unit in common use, however, is ten times smaller than this: the *decibel* (db). That is:

$$\text{Gain in decibels} = 10\log_{10}\left(\frac{P_o}{P_i}\right) = 20\log_{10}\left(\frac{V_o}{V_i}\right) \tag{8.23}$$

In this way certain gains of less than a decade appear as integral numbers of decibels. A range of decibel values is given in Table 8.1. The decibel scale proves to be so useful that it is common practice to disregard whether the resistors in which the power is developed are in fact equal. Equation (8.23) is thus used as an operational definition of gain in decibels. Plotting gain in decibels versus frequency on a logarithmic scale we produce what is known as a *Bode plot*, named after the engineer who popularized this procedure.

Example 8.6 Apart from the ratio of output power to input power for a system, another ratio of extreme interest is that of the signal power to the noise power. Indeed, achieving a high system gain could prove quite futile if one is dealing with a signal of poor signal-to-noise ratio since *both* signal and noise would be amplified producing no overall benefit. Signal-to-noise ratios are also usually given in decibels. As an example, the signal-to-noise ratio for the compact disc (CD) is generally quoted at about 80 db while magnetic tape performs at 50 db. In the former case this means that when the induced voltage and current fluctuations are translated into powers the (mean) noise power is 8 orders of magnitude below the maximum signal power while in the latter case it is only 5 orders of magnitude down. However, it should be borne in mind that in terms of the voltage or current fluctuations it is only a factor of 4 orders of magnitude or 2.5 orders. Noise, by its very nature, is random and can only be discussed statistically, so while the voltage corresponding to the mean noise power, the mean square noise voltage, may be 4 orders of magnitude below the maximum signal there is a finite probability that some noise excursions will approach the signal level. It requires a very high signal-to-noise ratio indeed to ensure a silent background in the quiet passages.

$\dfrac{V_o}{V_i}$	$\dfrac{P_o}{P_i}$	db
1000	10^6	60
100	10^4	40
10	10^2	20
2	4	6
$\sqrt{2}$	2	3
1	1	0
$\dfrac{1}{\sqrt{2}}$	$\dfrac{1}{2}$	-3
$\dfrac{1}{2}$	$\dfrac{1}{4}$	-6

Table 8.1

8.6 First-order Filters

As a reminder, let us say that by 'first-order' we mean a network for which s appears to order 1 in the denominator of $H(s)$ and consequently $j\omega$ appears to order 1 in the denominator of $H(j\omega)$. Both follow from the fact that the system differential equation is first-order. The circuits which follow are the minimum configurations to achieve the particular responses and would serve as models for other networks the internal details of which are not known but which display that response.

Example 8.7 Consider the circuits in Figure 8.17 which are taken together because, as we show, they are identical in respect of the fact that they pass low

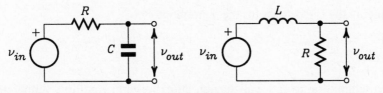

Figure 8.17

frequencies but attenuate high frequencies: they are *low-pass filters*. The frequency response function is obtained for each circuit by taking the ratio of output phasor to input phasor, as in Chapter 5, but now using the steady a.c. impedance rather than the generalized impedance for each component. Treating each circuit as an impedance potential divider, again as in Chapter 5, we may write down the frequency response functions by inspection:

$$H(j\omega)_{RC} = \frac{1/j\omega C}{R+1/j\omega C}, \quad H(j\omega)_{LR} = \frac{R}{R+j\omega L}$$

that is $\quad H(j\omega)_{RC} = \dfrac{1}{1+j\omega CR}, \quad H(j\omega)_{LR} = \dfrac{1}{1+j\omega L/R}$

The frequency response function for each of the filters in the above example may be written in the form:

$$H(j\omega) = G\frac{1}{1+j\omega/\omega_0} \tag{8.24}$$

where $\omega_0 = 1/CR$ for the RC circuit, $\omega_0 = R/L$ for the LR circuit and $G = 1$. The simple RC and LR filters shown are archetypes for this class of response which can, in general, be realized in an indefinite number of ways when $G \le 1$ for any passive circuit and will take any finite value for a (stable) active circuit.

The frequency response function is a complex quantity and may be written in the Euler form:

$$H(j\omega) = |H(j\omega)|e^{j\delta} \tag{8.25}$$

Expressing it as a ratio of phasors:

$$H(j\omega) = \frac{\mathbf{y}}{\mathbf{x}} = \frac{y_0 e^{j\Phi_o}}{x_0 e^{j\Phi_i}}$$

where Φ_o and Φ_i are the initial phases of the output and input phasors respectively. We see that the modulus of $H(j\omega)$ is the ratio of output amplitude to input amplitude (y/x), the gain, while the argument of $H(j\omega)$ is the phase difference $(\Phi_o - \Phi_i)$ between output and input. Similiar statements were made earlier about the modulus and argument of the steady a.c. impedance. The frequency response functions for first-order low-pass circuits have been obtained in the form:

$$H(j\omega) = \frac{1}{a+jb}$$

or, introducing the Euler form for the denominator:

$$H(j\omega) = \frac{1}{re^{j\delta'}}$$

There is a temptation to rationalize such an expression (multiply numerator and denominator by the complex conjugate of the denominator) to find the modulus and

$$\omega \to 0 \qquad |H(j\omega)| \to 1 \qquad \delta \to 0$$

$$\omega = \omega_0 \qquad |H(j\omega)| = 1/\sqrt{2} \qquad \delta = -\pi/4$$

$$\omega \to \infty \qquad |H(j\omega)| \to \omega_0/\omega \qquad \delta \to -\pi/2$$

Table 8.2 First-order low-pass filter characteristics

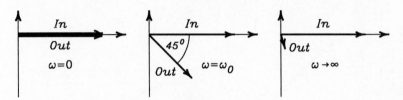

Figure 8.18

argument but this is not necessary. The modulus is simply $1/r$ with $r = \sqrt{(a^2 + b^2)}$ and the argument $\delta = -\delta'$. Consequently we find for the circuits of Figure 8.17:

$$|H(j\omega)| = \frac{1}{\sqrt{1 + (\omega/\omega_0)^2}} \quad \text{and} \quad \tan(-\delta) = \omega/\omega_0$$

The general features of the frequency dependence of the gain and the phase change may be seen by examining the behaviour as $\omega \to 0$ and as $\omega \to \infty$. The results are given in Table 8.2 together with the intermediate case $\omega = \omega_0$ which leads to a particularly useful way of characterizing a filter. For simplicity we have taken $G = 1$. The same information can be given in a series of phasor diagrams as in Figure 8.18. The value of such simple pictorial representations to reinforce our understanding should not be discounted.

The detailed frequency dependence of the gain is shown in Figure 8.19 together with that for the phase change. On this occasion the variation on both linear and logarithmic scales is illustrated. In the latter case a decibel scale is provided for the gain, so producing a Bode plot. For the low and high frequency asymptotes of the gain given in Table 8.2 the corresponding expressions referred to purely logarithmic scales are:

$$\log_{10}|H(j\omega)| = 0$$
$$\text{and} \quad \log_{10}|H(j\omega)| = \log_{10}(\omega_0/\omega)$$

The latter indicates that:

$$\log_{10}|H(j\omega)| \propto -\log_{10}\omega$$

which describes a straight line of slope -1. That is, for each decade (ten-fold increase) of frequency the gain decreases by a decade. On a Bode plot that is a 20 db decrease. It is at least as usual to refer to a doubling or halving of frequency

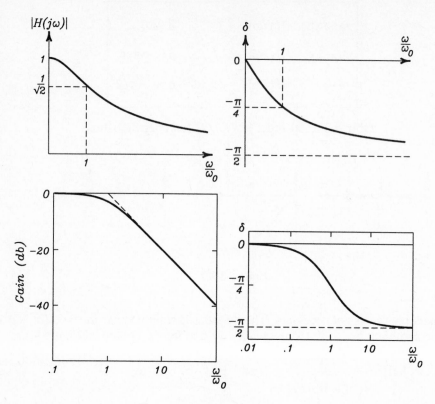

Figure 8.19

(an octave) as it is to use the frequency decade. In this case there is a 6 db decrease in gain. Specification in these terms, decibels per octave or decade, is part of the almost private language of electronic engineers.

The two straight line asymptotes on the logarithmic plot intersect where:

$$\log_{10}(\omega/\omega_0) = 0$$

that is, where $\omega/\omega_0 = 1$ or $\omega = \omega_0$

It follows from this that ω_0 is usually known as the 'corner frequency' (the frequency at the 'corner' of the asymptotes). It was for this reason that we included ω_0 in Table 8.2. At ω_0 the gain is seen to be $1/\sqrt{2}$ or, from Table 8.1, -3 db. Accordingly the gain at the corner frequency is also referred to as the 3 db point (minus sign understood). The asymptotes are often used as a rough representation of the frequency dependence but it must be emphasized that the actual characteristic follows a smooth curve from one asymptote to the other.

Example 8.8 Next let us examine the circuits of Figure 8.20, which are *high-pass filters*. In the same way as before we write down the frequency response function by inspection:

$$H(j\omega)_{RC} = \frac{R}{R+1/j\omega C}, \quad H(j\omega)_{LR} = \frac{j\omega L}{R+j\omega L}$$

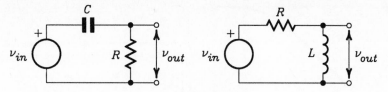

Figure 8.20

$\omega \to 0$	$\|H(j\omega)\| \to \omega/\omega_0$	$\delta \to \pi/2$
$\omega = \omega_0$	$\|H(j\omega)\| = 1/\sqrt{2}$	$\delta = \pi/4$
$\omega \to \infty$	$\|H(j\omega)\| \to 1$	$\delta \to 0$

Table 8.3 First-order high-pass filter characteristics

that is $$H(j\omega)_{RC} = \frac{j\omega CR}{1 + j\omega CR}, \quad H(j\omega)_{LR} = \frac{j\omega L/R}{1 + j\omega L/R}$$

In this instance the frequency response functions may be written in the general form:

$$H(j\omega) = G\frac{j\omega/\omega_0}{1 + j\omega/\omega_0} = G\frac{1}{1 - j\omega_0/\omega} \tag{8.26}$$

where ω_0 is again $1/CR$ for the RC circuit and R/L for the LR circuit. The second form of $H(j\omega)$ is rather more convenient for finding the modulus and argument (the gain and phase) which are

$$|H(j\omega)| = \frac{G}{\sqrt{1 + (\omega_0/\omega)^2}} \quad \text{and} \quad \tan(-\delta) = -\omega_0/\omega$$

Again for simplicity taking $G = 1$, the asymptotic values and the 3 db point are given in Table 8.3 and the Bode and phase plots are shown in Figure 8.21.

8.7   Second-order Band-pass Filters

In the same way that we issued a reminder in connection with first-order filters so we now remark that by second-order we mean a network for which the highest power of s in the denominator of $H(s)$ is 2. Consequently, $j\omega$ appears to a power no higher than 2 in the denominator of $H(j\omega)$. The behaviour of both denominators follows from the fact that the system differential equation is second-order. There is the possibility of more variants within the second-order class and it is appropriate to sub-divide their discussion, as we do now, drawing attention first to the feature for which they are best known, the band-pass characteristic.

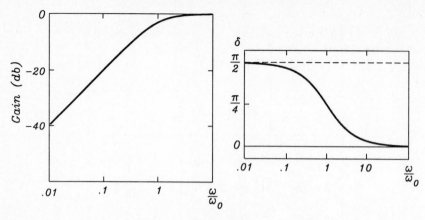

Figure 8.21

Example 8.9 Again consider two networks together, those depicted in Figure 8.22, because, as we shall see, they have the same general frequency characteristic. They are often described, rather colloquially, as the *series* resonance and the *parallel* resonance circuits. (It should be clear to which part of Figure 8.22 each of the descriptions refers.) Emphasizing the mere series or series/parallel connection of components disguises the fact that a number of distinct frequency responses follow from the precise placing of the input and output. We examine others in a later section. Making use of the driving-point impedances obtained earlier for series LC and parallel LC combinations, equations (8.15) and (8.16), we may treat the two networks as impedance potential dividers and write down the frequency response functions by inspection. Taking the series/parallel combination first:

$$H(j\omega)_{par} = \frac{\dfrac{L/C}{j(\omega L - 1/\omega C)}}{R + \dfrac{L/C}{j(\omega L - 1/\omega C)}}$$

Multiplying numerator and denominator by $j(\omega L - 1/\omega C)$ and at the same time dividing by R produces an expression generally similiar to that for the series LCR circuit which we can write down immediately by inspection:

$$H(j\omega)_{ser} = \frac{R}{R + j(\omega L - 1/\omega C)}, \quad H(j\omega)_{par} = \frac{L/CR}{L/CR + j(\omega L - 1/\omega C)}$$

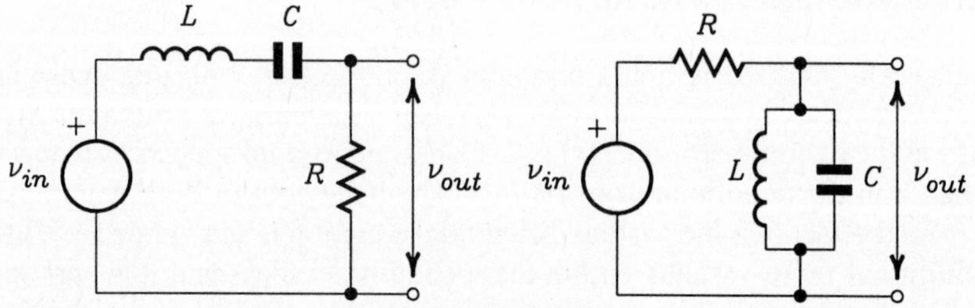

Figure 8.22

$$
\begin{array}{lll}
\omega \to 0 & |H(j\omega)| \to \dfrac{\omega}{\omega_0 Q} & \delta \to \dfrac{\pi}{2} \\[2mm]
\omega = \omega_0 & |H(j\omega)| = 1 & \delta = 0 \\[2mm]
\omega \to \infty & |H(j\omega)| \to \dfrac{\omega_0}{\omega Q} & \delta \to \dfrac{-\pi}{2}
\end{array}
$$

Table 8.4 Second-order band-pass filter characteristics

At this stage we may make certain general observations: for both circuits there is a frequency ω_0 for which $\omega_0 L = 1/\omega_0 C$, at this frequency the denominator is *wholly real*, therefore $H(j\omega)$ is real, the output and input are *in phase* and δ is zero. Further, the modulus of the denominator is a minimum so that $|H(j\omega)|$ is a maximum and indeed that value is 1. Obviously from the above $\omega_0 = 1/\sqrt{(LC)}$ and we have previously seen this (Chapter 4) as the frequency of natural oscillations of an LC combination. As indicated then, when such a circuit is stimulated near this natural frequency it will respond strongly, *resonate*, hence the term resonant circuits. At ω_0 the series circuit will appear simply as a shunt resistor to the input while the parallel LC series R will appear as a series resistor. For this reason also it follows that the transfer function must be unity at ω_0.

To render both expressions in a form with the denominator as a polynomial in $j\omega$ the numerator and denominator of both are multiplied by $j\omega C$:

$$
H(j\omega)_{ser} = \frac{j\omega CR}{1 + j\omega CR - \omega^2 LC}, \quad H(j\omega)_{par} = \frac{j\omega L/R}{1 + j\omega L/R - \omega^2 LC}
$$

The two transfer functions may be presented in a single, we might say standard, form by introducing ω_0 from above, the factor Q first mentioned in Chapter 5 and noting that $-\omega^2 = (j\omega)^2$:

$$
H(j\omega) = \frac{\dfrac{j\omega}{\omega_0 Q}}{1 + \dfrac{j\omega}{\omega_0 Q} - \dfrac{\omega^2}{\omega_0^2}} \tag{8.27}
$$

For the series circuit, $Q = 1/\omega_0 CR$ while for the series/parallel circuit $Q = R/\omega_0 L = \omega_0 CR$. In more general situations there will be an overall gain factor, G, but we omit this here so as not to obscure the main features of the filters.

The behaviour of the response function may be examined at certain specimen frequencies as for the first-order filters and the results are given in Table 8.4. The gain and phase plots (modulus and argument of $H(j\omega)$) are shown in Figure 8.23. An indication of the effect of changing Q is given in the figure.

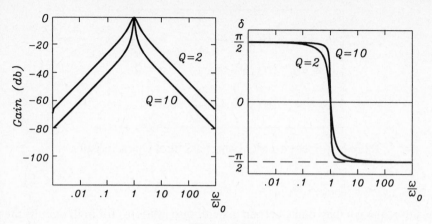

Figure 8.23

8.8 The Quality Factor, Q

A factor Q has now been introduced, somewhat mysteriously, on two occasions to allow denominators in second-order system functions and frequency response functions to be written in a 'standard form'. An account of the central role of this factor may not be delayed any longer and is made simple by the availability now of the second-order frequency response curve (Figure 8.24). At the 3 db points of the band-pass characteristic $|H(j\omega)| = 1/\sqrt{2}$ and so $|H(j\omega)|^2 = 1/2$. Now

$$|H(j\omega)|^2 = H(j\omega)H^*(j\omega)$$

where $H^*(j\omega)$ is the complex conjugate of $H(j\omega)$. It follows, using the standard form of the frequency response function, that:

$$|H(j\omega)|^2 = \frac{\dfrac{j\omega}{\omega_0 Q}}{\left(1 - \dfrac{\omega^2}{\omega_0^2}\right) + \dfrac{j\omega}{\omega_0 Q}} \cdot \frac{\dfrac{-j\omega}{\omega_0 Q}}{\left(1 - \dfrac{\omega^2}{\omega_0^2}\right) - \dfrac{j\omega}{\omega_0 Q}}$$

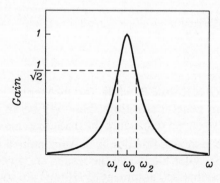

Figure 8.24

Therefore:

$$\frac{\left(\dfrac{\omega}{\omega_0 Q}\right)^2}{\left(1-\dfrac{\omega^2}{\omega_0^2}\right)^2+\left(\dfrac{\omega}{\omega_0 Q}\right)^2}=\frac{1}{2}$$

at the 3 db points where $|H(j\omega)|^2=\frac{1}{2}$. Multiplying out and collecting terms we obtain:

$$\left(1-\frac{\omega^2}{\omega_0^2}\right)^2-\left(\frac{\omega}{\omega_0 Q}\right)^2=0$$

which is a difference between two squares generating a product of two terms each of which separately may be zero; that is either:

$$1-\frac{\omega^2}{\omega_0^2}-\frac{\omega}{\omega_0 Q}=0 \quad \text{or} \quad 1-\frac{\omega^2}{\omega_0^2}+\frac{\omega}{\omega_0 Q}=0$$

These are two quadratic equations with roots:

$$\frac{\omega_1}{\omega_0}=-\frac{1/Q\pm\sqrt{1/Q^2+4}}{2},\quad \frac{\omega_2}{\omega_0}=-\frac{-1/Q\pm\sqrt{1/Q^2+4}}{2}$$

where ω_1 and ω_2 have been introduced as the frequencies at the two 3 db points. Taking only the positive roots:

$$\frac{\omega_2}{\omega_0}-\frac{\omega_1}{\omega_0}=\frac{1}{Q}$$

that is $\quad \dfrac{\omega_2-\omega_1}{\omega_0}=\dfrac{1}{Q}$

or $\quad \dfrac{\Delta\omega}{\omega_0}=\dfrac{1}{Q}$

where $\Delta\omega=\omega_2-\omega_1$. Hence

$$Q=\frac{\omega_0}{\Delta\omega} \tag{8.28}$$

The *smaller* the bandwidth of a response the larger is Q for the corresponding circuit and in that sense the higher is its 'quality'. For any particular circuit Q will be a certain function of component values. We have already seen two such cases for the two band-pass filters just analyzed.

Example 8.10 The radio receiver suggested in Example 8.1 as being able to isolate radio stations operating at about 100 MHz would require an effective Q for its tuned circuits of $100\times10^6/200\times10^3=500$. Such a high value could readily be obtained in professional or research equipment but places quite high demands on equipment designed for domestic purposes where temperature stability would prove a problem. Accordingly, the selectivity of the instrument is achieved by indirect means involving both transferring the modulating signal to a lower-frequency internal carrier (intermediate frequency techniques) and using circuits with feedback (built around the phase-locked-loop) which lock onto the required frequency.

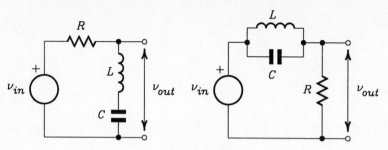

Figure 8.25

8.9   Second-order Notch, Low-pass and High-pass Filters

In the context of their use as filters the series resonance and parallel resonance circuits allow a number of different responses depending upon the placing of the input and output, as was suggested earlier. The most obvious variation on the band-pass configuration so far considered is to take the output across the other half of the impedance potential divider in each case, that is, to make the connections shown in Figure 8.25. The band-pass filter has been 'turned upside down' in each case, with a fairly predictable outcome for the responses, which by inspection are:

$$H(j\omega)_{ser} = \frac{Z_{ser}}{R+Z_{ser}}, \quad H(j\omega)_{par} = \frac{R}{R+Z_{par}}$$

where, for convenience in the algebra, we have introduced Z_{ser} and Z_{par} rather than the complete expressions given in equations (8.15) and (8.16). It follows that:

$$H(j\omega)_{ser} = 1 - \frac{R}{R+Z_{ser}}, \quad H(j\omega)_{par} = 1 - \frac{Z_{par}}{R+Z_{par}}$$

and in each case $H(j\omega) = 1 - H(j\omega)_{bandpass}$

The behaviour of the function at the usual specimen frequencies is given in Table 8.5. For a high Q (narrow bandwidth) circuit the high and low frequency values of the gain are very close to 1. The frequency dependence of the gain is shown in Figure 8.26 from which it should be clear why the filter is described as a line-reject or notch filter.

$\omega \to 0$	$|H(j\omega)| \to 1$	$\delta \to$	$\dfrac{-\pi}{2}$
$\omega = \omega_0$	$|H(j\omega)| = 0$	$\delta = 0$	
$\omega \to \infty$	$|H(j\omega)| \to 1$	$\delta \to$	$\dfrac{\pi}{2}$

Table 8.5 Notch filter characteristics

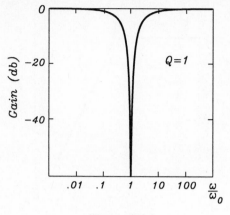

Figure 8.26

Example 8.11 A notch filter finds application in the removal of a particular un-wanted frequency, such as mains hum (50 Hz), from a signal. In such a case we have:

$$\frac{1}{\sqrt{LC}} = 100\pi$$

Choosing a nominal 1 μF capacitor, the value required for L is 10 H which is an impracticable value for any equipment considered to be reasonably portable. How-ever, an active circuit is available which will rotate the current and voltage phasors on the Argand diagram so that while provided with a capacitor it will appear as an inductor at its terminals. Such circuits are, not surprisingly, known as gyrators. In this way an effective 10 H can be obtained and a 50 Hz line-reject filter can be implemented.

In the case of the series resonance circuit we may also choose to take the output across either the capacitor (Figure 8.27(a)) or the inductor (Figure 8.27(b)). The general features of these circuits may be seen by considering the first as $\omega \to 0$, when the reactance of the inductor becomes negligible, and the second as $\omega \to \infty$, when the reactance of the capacitor becomes negligible. The former approximates to the RC low-pass filter, for which we expect the low frequency gain to be 1, and the latter to the RL high-pass filter, for which we expect the high frequency gain

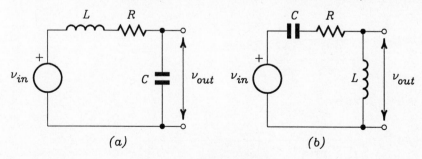

(a)   (b)

Figure 8.27

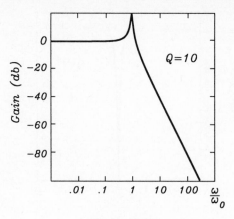

Figure 8.28

to be 1. Between low and high frequencies some evidence of the characteristic resonance peak will generally be seen, as shown in Figure 8.28. The reader is left to show that the expressions corresponding to these responses, for the second-order low-pass and second-order high-pass filters, are:

$$H(j\omega)_{lp} = \frac{1}{1 + \dfrac{j\omega}{\omega_0 Q} - \dfrac{\omega^2}{\omega_0^2}}, \quad H(j\omega)_{hp} = \frac{-\dfrac{\omega^2}{\omega_0^2}}{1 + \dfrac{j\omega}{\omega_0 Q} - \dfrac{\omega^2}{\omega_0^2}} \qquad (8.29)$$

A simple algebraic connection may be noticed between the expressions for the low-pass, band-pass and high-pass responses in that they are generated by using for the numerator either the zeroth-order term from the denominator (low-pass), the first-order term (band-pass) or the second-order term (high-pass). Much attention is given to the band-pass feature of second-order filters but the low-pass and high-pass characteristics we have just described also have great utility. They are usually exploited in situations where resonance has been suppressed by arranging for the Q of the circuit to be small.

Example 8.12 As an example of a low Q second-order low-pass filter we take the case of $Q = 1/\sqrt{2}$:

$$H(j\omega)_{lp} = \frac{1}{1 + \dfrac{j\sqrt{2}\omega}{\omega_0} - \dfrac{\omega^2}{\omega_0^2}}$$

The gain is then given by:

$$|H(j\omega)|_{lp}^2 = \frac{1}{\left(1 - \dfrac{\omega^2}{\omega_0^2}\right)^2 + 2\left(\dfrac{\omega}{\omega_0}\right)^2} = \frac{1}{1 + \left(\dfrac{\omega}{\omega_0}\right)^4}$$

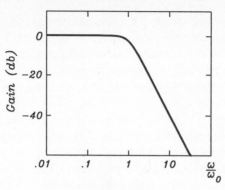

Figure 8.29

We may write the gain as the special case of a more general formula:

$$|H(j\omega)|_{lp} = \frac{1}{\sqrt{1 + \left(\dfrac{\omega}{\omega_0}\right)^{2n}}} \qquad (8.30)$$

which is the so-called Butterworth response of order n and where, in this case, $n = 2$. The Butterworth response (Figure 8.29) has the feature of being maximally flat in the pass band. It cuts off at high frequency at 12 db per octave, twice as fast as a first-order filter. The sharper cut-off is exactly the reason why a low-pass filter of high order would be chosen in preference to one of lower order. In critical applications filters of very high order will be used, often being realized by cascading, with suitable buffering, second-order modules such as we have just described.

Summary

The emphasis has moved in this chapter from more purely analytical matters closer to the *design* of circuits, completely in line with the expectation expressed in Chapter 1 that analysis must precede synthesis. We have previously seen the virtue of representing components by their impedances. Subsequently we agreed that understanding the response of networks to steady a.c. allows us to describe their behaviour for a wide variety of other stimuli including inputs which are aperiodic, such as an impulse. In very many cases of practical interest it is appropriate to discuss the operation of a network with just a steady a.c. input. Here we have combined the idea of component impedance and operation at steady a.c. to survey some of the simplest building blocks of electronic circuits.

We have continued to make full use of the phasor (complex exponential) notation bearing in mind that in practical terms the magnitude of such complex numbers is the *amplitude* (of a voltage or current) and the argument is the *initial phase* (of the voltage or current). Applying a steady a.c. source of frequency ω, for which the time-dependence is described by the factor $e^{j\omega t}$, to each of the components R, L and

C in turn, we deduced that the steady a.c. impedance of a resistor is simply its resistance, R, while for an inductor it is $j\omega L$ and for a capacitor, $1/j\omega C$, or $j(-1/\omega C)$. For combinations of such components the total impedance Z is $R + jX$, where R is the real, or *resistive*, part and X is the imaginary, or *reactive*, part. It follows that an inductor in isolation is wholly reactive with reactance $X_L = \omega L$; likewise for a capacitor the reactance $X_c = -1/\omega C$. Depending on how a particular group of components is connected together, the resistive part of the net impedance may be a combination of pure resistance and reactance; similarly the reactive part may be a combination of resistance and reactance. Only in the simplest case of a series combination of resistors, inductors and capacitors is the real part of the impedance pure resistance and the imaginary part pure reactance. When the impedance is presented in its Euler form, $|Z|e^{j\delta}$, the argument δ is the difference in the initial phases, or simply the *phase difference*, of the phasors for the voltage across the impedance and the current through the impedance. For the case of an inductor or a capacitor in isolation it is possibly simpler to remember that multiplication by j corresponds to a rotation of 90° in the Argand diagram and so from the wholly imaginary impedances of these components we deduce that the current phasor is rotated 90° with respect to the voltage phasor in both cases, current leading voltage for the capacitor and lagging for the inductor.

When components with mutual inductance (transformers) are present the effect of impedance in the secondary circuit can be translated to the primary circuit, and the transformer thereafter considered to have zero coupling (no mutual inductance). For the common case of transformers with closely coupled windings of high inductance, so-called *ideal* transformers, commonly used for the transfer of power, the *reflected impedance* in the primary is the impedance in the secondary circuit Z_2, reduced by the square of the turns ratio, secondary to primary, n, that is Z_2/n^2.

Once again exploiting the properties of the potential divider, but now with each arm a steady a.c. impedance, we could write down, by inspection in simple cases, the ratio of output to input amplitudes for steady a.c., $H(j\omega)$. That ratio we have previously called the *frequency response*. We then classified networks of resistors, capacitors and inductors according the power of $j\omega$ in the denominator of $H(j\omega)$. The power of $j\omega$ is the *order* of the circuit and reflects the order of the differential equation which fundamentally describes the circuit (but which we may now avoid by the use of impedances). A first-order circuit includes just one energy storage component, capacitor or inductor, a second-order resonant circuit must include both a capacitor and an inductor. Higher-order circuits were not examined but may be produced by cascading first- and second-order modules. Circuits were further classified in terms of their overall frequency response characteristics, as *low-pass, high-pass* and *band-pass* and generally described as *filters*, acknowledging the fact that any circuit will, intentionally or otherwise, change, or *filter*, the frequency content of the input.

The Q factor, having been introduced in a formal sense much earlier to express the coefficient of the first-order term in the denominator of both $H(s)$ and $H(j\omega)$ for second-order circuits, was seen to be directly related to the width of the resonant peak in the response of such circuits. High Q circuits are characterized by a narrow resonant peak, while for a low Q circuit the peak is broad. Second-order low-pass and high-pass circuits with a low Q factor (even lower than unity) were seen to have a particular use in that, while the resonant peak is entirely absent, the separation between the pass-band and the stop-band is much more abrupt than for a first-order filter.

Recognizing that in practical terms one is interested in how the response of a system changes with each order-of-magnitude change in frequency rather than with the change on a linear scale, it is appropriate to work with a *logarithmic* frequency scale. Similarly we do not require the same detail in the range from 10% to 100% of the output of a system as we do in the range 1% to 10% or 0.1% to 1% and so on. Again a logarithmic scale makes more sense. More importantly it is the *power* at the output as it relates to the input that is most relevant. The logarithm of the power ratio (output to input) is expressed in *bels*, but to round-up common fractional power ratios to whole numbers a unit which is ten times smaller, the *decibel* (abbreviated to db), is considered more convenient. Plotting the gain of a system in decibels versus frequency on a logarithmic scale produces what is termed a *Bode plot*, after the engineer who promoted this procedure.

Problems

8.1 Translate the gains for each of the blocks shown in Figure 8.30 to decibels and hence find the overall gain of the system in decibels. Convert the gain in decibels to a numerical gain and compare it with the value which may be obtained directly by inspection.

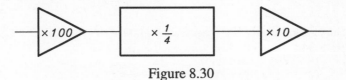

Figure 8.30

8.2 Calculate the driving-point impedance for the LCR combination in Figure 8.31 when the angular frequency of the source is $10\,000\,\text{rad sec}^{-1}$. Hence find the amplitude of the current i when the voltage source v has unit amplitude, also find the phase of i relative to v.

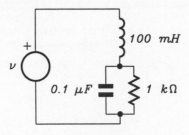

Figure 8.31

8.3 Find the turns ratio of the ideal transformer required to achieve maximum power transfer in the circuit in Figure 8.32.

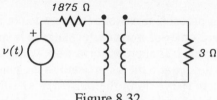

Figure 8.32

8.4 Show that for the circuit in Figure 8.33 the modulus of the ratio of the voltage between A and B to the source voltage is $\frac{1}{2}$ at all frequencies. Obtain an expression for the frequency at which the two voltages are 90° out-of-phase.

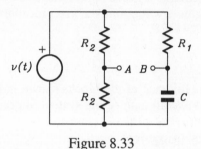

Figure 8.33

8.5 Obtain an expression for the frequency response, $H(j\omega)$, of the circuit shown in Figure 8.34.

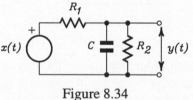

Figure 8.34

8.6 Calculate the resonance frequency and the Q factor for the network in Figure 8.35.

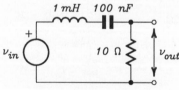

Figure 8.35

8.7 By differentiating the modulus of the frequency response function for the second-order low-pass filter, equation (8.29), to find the turning point, obtain an expression for the frequency of maximum response.

9

Power Dissipation and Energy Storage in Networks

9.1 Power Dissipation in Resistive Networks

The simplest ideas regarding power in networks were taken for granted in the preceding chapters. They derive from the fundamental concept of the work done in transporting a charge around a closed circuit. For a d.c. source in a resistive circuit (Figure 9.1(a)), the work done to transport a charge dq around the loop is $dW = vdq$. The power, the rate of doing work, is

$$\frac{dW}{dt} = v\frac{dq}{dt} = v\,i$$

Using Ohm's law we obtain the already familiar result

$$p_{dc} = \frac{v^2}{R} = i^2 R \tag{9.1}$$

For a steady a.c. source in a resistive circuit (Figure 9.1(b)), we may only write down the rate of doing work at a particular instant, the *instantaneous power*:

$$p(t) = \frac{v^2(t)}{R} = i^2(t)R \tag{9.2}$$

The complex exponential formalism offers no advantage at this stage and we express the steady a.c. voltage in simple trigonometric form:

$$v(t) = v_0 \cos(\omega t + \Phi)$$

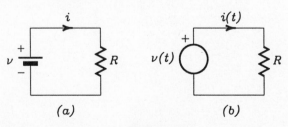

(a)   (b)

Figure 9.1

where Φ is the initial phase. A similar expression would obtain for the current. Working just in terms of the voltage across the resistor:

$$p(t) = \frac{v_0^2 \cos^2(\omega t + \Phi)}{R}$$

$$= \frac{v_0^2}{R} \frac{1}{2} \left[1 + \cos 2(\omega t + \Phi)\right] \tag{9.3}$$

By virtue of the presence of energy storage elements, inductors and capacitors, the *instantaneous* power delivered by a network to a load may be very different from the power supplied to the network by a source. The only practicable way to make a comparison between the power supplied and the power delivered in such circumstances is to discuss the *mean* power. Often only a vague distinction is drawn between the mean and the *average*. The former entails summing (integrating) over a time which approaches infinity and dividing by that time to obtain an 'expected' value:

$$\overline{p(t)} = \lim_{T \to \infty} \frac{1}{T} \int_{-T/2}^{T/2} p(t)\, dt$$

The bar above the quantity is the usual convention for indicating the mean. For a strictly periodic function in which each period conveys the same information as every other, the limit may be dropped and the integral taken over just one period. In this case and any other in which a finite interval is considered, the term 'average' is often used in preference to 'mean'.

Continuing to work just in terms of the voltage across the resistor we take the mean of equation (9.2) to obtain for the mean power:

$$\overline{p(t)} = \frac{\overline{v^2(t)}}{R} \tag{9.4a}$$

For the steady a.c. source introduced above we may use equation (9.3) to obtain:

$$\overline{p(t)} = \frac{v_0^2}{R} \frac{1}{2} \left[\overline{1 + \cos 2(\omega t + \Phi)}\right] \tag{9.4b}$$

Now the mean of any sinusoidal (or cosinusoidal) function is zero. The result follows from the fact that successive half-cycles cancel in the integration. Therefore from equation (9.4b) we obtain:

$$\overline{p(t)} = \frac{1}{2} \frac{v_0^2}{R} \tag{9.5}$$

In terms of the current:

$$\overline{p(t)} = \frac{1}{2} i_0^2 R \tag{9.6}$$

Equations (9.5) and (9.6) resemble (9.2) for the d.c. case except that now one half the source amplitude squared appears instead of the magnitude of the d.c.

source. The resemblance prompts us to define a d.c. source which would deliver the same mean power as the a.c. source:

$$\overline{p(t)} = \frac{\overline{v^2(t)}}{R} = \frac{V^2}{R}$$

Clearly the magnitude of the d.c. source, V, must be the square root of the mean of the square of the value of the a.c. source, the so-called root-mean-square, or r.m.s., value. Identifying this particular value by means of a subscript, V_{rms}, we see from equation (9.5) that it is related to the amplitude of the a.c. source thus:

$$V_{rms}^2 = \frac{1}{2} v_0^2$$

$$\text{or} \quad V_{rms} = \frac{v_0}{\sqrt{2}} \tag{9.7}$$

Similarly for the current

$$I_{rms} = \frac{i_0}{\sqrt{2}} \tag{9.8}$$

Example 9.1 The supply-line (mains) voltage for domestic and much industrial use in the UK is 230 V r.m.s. It is worth noting that the *peak* voltage associated with this is:

$$V_{peak} = \sqrt{2}\,230 = 325 \text{ V}$$

It is also worth noting that by its very nature the polarity of the supply reverses each half-cycle and a very simple collection of components (a 'voltage doubler' comprising two rectifiers and two capacitors) can be employed to take advantage of this by charging one capacitor on one half-cycle, the other on the next half-cycle and so realize the *peak-to-peak* voltage of about 650 V at the output. Such easily realized lethal voltages are the reason for the stress placed on the safe handling of mains operated equipment.

9.2 Power Dissipation in Networks with Reactive Components

In networks containing the energy storage elements L and C we have seen (section 8.2) that the voltage and current phasors will, in general, not be in phase. The ratio of phasors is the impedance. Rearranging equation (8.6) we obtain:

$$i_0 e^{j\Phi_1} = \frac{v_0}{|Z|} e^{j(\Phi_2 - \delta)}$$

showing that for positive δ the current phasor lags the voltage phasor (by δ). The usual convention is to refer the current phasor to the voltage phasor in this way. The phasor diagram for a current of zero initial phase would be as in Figure 9.2(a). The angle between the phasors is, of course, identical to the basal angle of the associated impedance triangle (Figure 9.2(b)). Once again adopting trigonometric

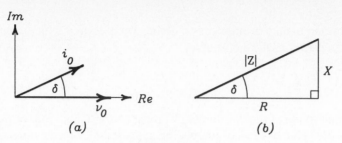

Figure 9.2

notation to describe steady a.c. voltages and currents, we may expand the general expression for the instantaneous power in a circuit containing reactive components:

$$p(t) = v(t)\, i(t)$$
to obtain
$$p(t) = v_0 \cos(\omega t + \Phi_2)\, i_0 \cos(\omega t + \Phi_1)$$
$$= v_0 \cos(\omega t + \Phi_2)\, i_0 \cos(\omega t + \Phi_2 - \delta)$$

There is no loss of generality, and it considerably simplifies the algebra, in taking the initial phase of the voltage to be zero. Then:

$$
\begin{aligned}
p(t) &= v_0 \cos \omega t\, i_0 \cos(\omega t - \delta) \\
&= v_0 i_0 \cos \omega t\, [\cos \omega t \cos \delta + \sin \omega t \sin \delta] \\
&= v_0 i_0 \cos^2 \omega t \cos \delta + v_0 i_0 \sin \omega t \cos \omega t \sin \delta \\
&= \frac{1}{2} v_0 i_0 [1 + \cos 2\omega t] \cos \delta + \frac{1}{2} v_0 i_0 \sin 2\omega t \sin \delta
\end{aligned}
\tag{9.9}
$$

As for the resistive circuit, so our interest for the circuit with reactive components will be in the mean power. Again using the fact that sinusoidal functions have zero mean we find:

$$\overline{p(t)} = \frac{1}{2} v_0 i_0 \cos \delta$$

Convention has it that the mean power is to be identified by $\mathcal{P}$. Introducing the definitions of r.m.s. voltage and current (equations (9.7) and (9.8)) we find:

$$\mathcal{P} = V_{rms} I_{rms} \cos \delta \tag{9.10}$$

The mean power is therefore not simply the VI product but is determined very much by $\cos \delta$, called the *power factor*.

It is possible to have very large currents out of phase with the voltage so that only a very small power is developed in a circuit. An extreme example is shown in Figure 9.3. There the voltage and current are 90° out of phase so that $\cos \delta = 0$. The reader will readily appreciate that, whatever the precise value of the voltage–current product integrated over the first quarter-cycle, the opposite sign will be obtained in the next quarter-cycle because one function changes sign. The same is true through-out all succeeding pairs of quarter-cycles so that when summed to infinity the total energy is zero and the mean power is therefore zero. The very real difficulty with this situation arises in the case of domestic and industrial power supply. While

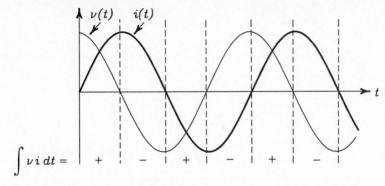

$$\int v i \, dt = \quad | \quad + \quad | \quad - \quad | \quad + \quad | \quad - \quad | \quad + \quad | \quad - \quad |$$

Figure 9.3

no, or very little power may be being developed in the user's equipment, the large out-of-phase currents must be delivered over the supply lines, where power, $I_{rms}^2 R$, *will* be dissipated (R is the resistance of the power lines). The supply company is therefore dissipating power in its own equipment and raising no revenue from the consumer. Understandably the power companies insist that consumers run their equipment at as close to unity power factor as possible. The procedure is called *power factor correction*. The usual situation with an industrial user is that there is motor-driven machinery presenting a large inductive load and the user must provide a large capacitor locally to compensate it.

To effect power factor correction one requires, in addition to the uncorrected power factor, information as to whether the load is inductive or capacitative and so whether the voltage leads the current or vice versa. That information is not conveyed by the sign of the power factor ($\cos \delta$) which is positive for both positive and negative δ. One must give the value of $\cos \delta$ and state explicitly 'lagging' or 'leading' referring to the phase of the current with respect to the voltage. In this connection the mnemonic CIVIL is often found quite helpful: Capacitor; current (I) leads voltage (V) and so on.

Example 9.2 As an illustration of the procedure we suppose that there is a requirement to provide *complete* power factor correction at 50 Hz for a load impedance of $(100 + j100)$ ohms (the supply company will usually require a minimum of 0.95). The impedance triangle is shown in Figure 9.4(a) from which it will be seen that $\cos \delta = 1/\sqrt{2}$ so that $\delta = 45°$. We would expect to connect the power factor correction component at a sub-station or supply distribution board as a shunt, that is in parallel with the supply. As we saw in Chapter 8, it may be easier when combining impedances in parallel to work with their admittances. The conductance and the susceptance respectively for the specified load are, using equation (8.8):

$$G = \frac{100}{100^2 + 100^2} = 0.5 \times 10^{-2} \text{ S}$$

$$B = -\frac{100}{100^2 + 100^2} = -0.5 \times 10^{-2} \text{ S}$$

and these are shown in an admittance triangle in Figure 9.4(b). If we can reduce the net susceptance to zero then the load seen by the supply will be pure conductance

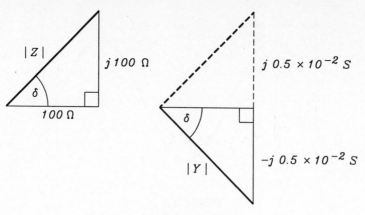

Figure 9.4

which equates to pure resistance with unity power factor. How to achieve this may easily be seen by referring to the Argand diagram depicting the admittance and a detailed algebraic calculation is not necessary. A susceptance of $+0.5 \times 10^{-2}$ will cancel that of the load. Such a susceptance equates to a component impedance:

$$Z = \frac{1}{j0.5 \times 10^{-2}} = j(-200)$$

which is a capacitive reactance. Then:

$$\frac{1}{\omega C} = 200 \quad \text{and with} \quad \omega = 100\pi$$

$$C \approx 16\mu F$$

We now return briefly to a restatement of the instantaneous power, equation (9.9). The symbol $\mathcal{P}$ was introduced to signify the mean power (equation (9.10)), and will now serve as a coefficient in the first term of equation (9.9). As part of the same convention we use the symbol Q for the coefficient of the time-dependent part of the second term:

$$Q = \frac{1}{2}v_0 \, i_0 \sin\delta$$

There is an unfortunate clash here with the symbol used for the quality factor, Q. We hope to distinguish the two by different type-styles. Using the definitions of r.m.s. voltage and current (equations (9.7) and (9.8)):

$$Q = V_{rms} \, I_{rms} \sin\delta$$

and is described as the reactive power. In terms of $\mathcal{P}$ and Q we may now write the instantaneous power:

$$p(t) = \mathcal{P}[1 + \cos 2\omega t] + Q\sin 2\omega t$$

The impedance has been defined in the foregoing as a ratio of phasors so that $|Z| = v_0/i_0$ where v_0 and i_0 are the voltage and current amplitudes. It should be clear from the uniform way in which r.m.s. values are related to amplitudes that we may equally well write $|Z| = V_{rms}/I_{rms}$. It is straightforward to see from the impedance triangle that $\cos\delta = R/|Z|$ and $\sin\delta = X/|Z|$. Using these simple relations a number of other forms of expression for the mean and reactive powers may be developed:

$$\mathcal{P} = |Z|I_{rms}I_{rms}\frac{R}{|Z|} \qquad\qquad Q = |Z|I_{rms}I_{rms}\frac{X}{|Z|}$$
$$\phantom{\mathcal{P}} = I_{rms}^2 R \qquad\qquad\qquad\qquad = I_{rms}^2 X$$

We see that the mean power is in one of the forms familiar to us in the case of d.c. and the reactive power is expressed in a similar way but using the reactance. However, we may alternatively find $\mathcal{P}$ and Q in terms of the voltage:

$$\mathcal{P} = V_{rms}\frac{V_{rms}}{|Z|}\frac{R}{|Z|} \qquad\qquad Q = V_{rms}\frac{V_{rms}}{|Z|}\frac{X}{|Z|}$$

in which case we see that $\mathcal{P} \neq V_{rms}^2/R$ unless $|Z| = R$, that is, unless $X = 0$. Similarly $Q \neq V_{rms}^2/X$ unless $|Z| = X$, that is, $R = 0$. Only one of the expressions familiar in the case of d.c. may therefore be translated to find the powers in the a.c. case.

9.3   The Complex Power

Returning to the use of the complex exponential (phasor) notation for voltages and currents we may develop a connection between the mean and the reactive powers in terms of a third quantity, the *complex power*. As with voltages and currents, a power cannot in fact be complex. It is, as with phasors, a useful device and allows us to draw on an Argand diagram a so-called power triangle which is geometrically similar to the impedance triangle discussed earlier. We approach this new quantity by first noting that where the quotient of voltage and current phasors is the impedance $Z = |Z|e^{j\delta}$, so that the current phasor lags the voltage phasor by δ, the product of phasors is:

$$\mathbf{vi} = v_0 e^{j\Phi} i_0 e^{j(\Phi-\delta)}$$
$$\phantom{\mathbf{vi}} = v_0 i_0 e^{j(2\Phi-\delta)}$$

The argument of the product bears no simple relationship to the argument of the impedance. However, if instead we take the product of one phasor and the complex conjugate of the other:

$$\mathbf{vi}^* = v_0 e^{j\Phi} i_0 e^{-j(\Phi-\delta)}$$
$$\phantom{\mathbf{vi}^*} = v_0 i_0 e^{j\delta}$$

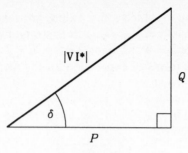

Figure 9.5

we have a complex quantity the argument of which *is* the same as that of the impedance. The same result may be expressed in terms of r.m.s. quantities by taking half of the above product, that is dividing each phasor and amplitude by $\sqrt{2}$:

$$\mathbf{VI}^* = V_{rms}\, I_{rms} e^{j\delta}$$

where $\mathbf{V}$ and $\mathbf{I}$ are new phasors the moduli of which are r.m.s. values rather than amplitudes. Expanding the Euler expression:

$$\begin{aligned} \mathbf{VI}^* &= V_{rms}\, I_{rms} \cos\delta + j V_{rms}\, I_{rms} \sin\delta \\ &= \mathcal{P} + jQ \end{aligned}$$

$\mathbf{VI}^*$ is the complex power and we see that its real part is the real (the mean) power while its imaginary part is the reactive power. The magnitude of the complex power is $V_{rms}\, I_{rms}$ and is termed the 'apparent power'. It is quoted as a volt-amps product (VA) and is generally greater than the real power by virtue of the power factor of the circuit. The reactive power is quoted in volt-amps reactive (VAR). Only the real power is given in watts. In any particular network containing resistors, capacitors and inductors, the real power is simply the sum of the powers dissipated in the resistors, as it would be if the network contained only resistors. In a similar way, the reactive power is the sum of the reactive powers in the reactive components. As with any other complex quantity the complex power may be depicted on an Argand diagram (Figure 9.5). The triangle formed by the modulus with the real and imaginary parts is geometrically similar to the triangle representing the impedance in which the power is being developed. The object of power factor correction can be seen as reducing the reactive power to a minimum which clearly entails reducing the reactance to a minimum.

9.4 Stored Energy in an LCR Circuit

On a number of occasions we have identified inductors and capacitors as *energy storage elements* but have concerned ourselves primarily with the consequences of their inclusion in circuits for the relative phases of voltages and currents. Now, by examining explicitly the energy stored in a circuit and its relation to the power dissipation we may develop a new perspective on what we have previously called the quality factor or Q. Any of the LCR circuits studied in Chapter 8 would suffice for this exercise. We choose probably the most familiar, the series LCR circuit configured as a band-pass filter (Figure 9.6). At any instant the energy stored in the inductor is $\frac{1}{2}Li^2$ while the energy stored in the capacitor is $\frac{1}{2}Cv^2$. Of course, both i and v are functions of time. Comparison with the results of Chapter 8 will be possible by confining our attention to steady a.c. Working once again with simple trigonometric expressions for current and voltage we take the current in the circuit to be $i_0 \cos\omega t$, also for brevity taking the initial phase to be zero. Then the energy stored in the inductor is:

$$W_L = \frac{1}{2}Li_0^2 \cos^2\omega t$$

The voltage across the capacitor will be 90° out of phase with the current so that the energy stored in the capacitor is:

$$
\begin{aligned}
W_C &= \frac{1}{2}Cv_0^2 \cos^2(\omega t - \pi/2) \\
&= \frac{1}{2}Cv_0^2 \sin^2\omega t
\end{aligned}
$$

The amplitude of the voltage across the capacitor, v_0, is related to the current in the circuit by the magnitude of the impedance of that component $(1/\omega C)$ so that

$$W_C = \frac{1}{2}C(1/\omega C)^2 i_0^2 \sin^2\omega t$$

At resonance in particular (the frequency of maximum response ω_0) we know that $1/\omega_0 C = \omega_0 L$. Therefore

$$W_C = \frac{1}{2}Li_0^2 \sin^2\omega_0 t$$

At resonance, therefore, the energy stored in the inductor and the capacitor are oscillatory functions of equal amplitude but 90° out of phase (when one is maximum the other is zero). The total energy stored in the circuit *at resonance* is

$$
\begin{aligned}
W_{TOTAL} &= \frac{1}{2}Li_0^2 \cos^2\omega_0 t + \frac{1}{2}Li_0^2 \sin^2\omega_0 t \\
&= \frac{1}{2}Li_0^2 = LI_{rms}^2
\end{aligned}
$$

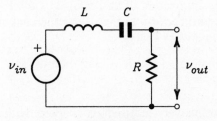

Figure 9.6

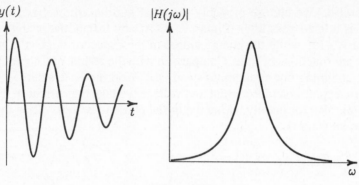

Figure 9.7

Now at resonance the circuit appears as a pure resistance, as we saw in Chapter 8, consequently the power dissipated is simply RI_{rms}^2. We may therefore write the ratio

$$\left[\frac{\text{Total stored energy}}{\text{Power dissipation}}\right]_{\omega_0} = \frac{LI_{rms}^2}{RI_{rms}^2} = \frac{L}{R}$$

Multiplying both sides of the expression by ω_0 we produce as the right-hand side $\omega_0 L/R$ which is what we defined to be the Q of this particular circuit. Writing ω_0 in an alternative form as $2\pi/T$ where T is the period of the oscillation and then recognizing that power dissipation $\times T$ is the energy lost per cycle, we find that

$$Q = 2\pi \left[\frac{\text{Total stored energy}}{\text{Energy lost per cycle}}\right]_{\omega_0}$$

It appears, therefore, that a low Q circuit, apart from having a broader frequency response characteristic, is also one which dissipates a higher proportion of its stored energy than does a high Q circuit which has a narrow frequency response. The two apparently disparate properties can be brought together when we realize that one property describes a behaviour in the frequency domain whereas the other relates to the time domain. The rate of energy loss in a circuit will dictate how heavily damped, or otherwise, are the free oscillations of the circuit. The Fourier transform of the damped oscillations (Figure 9.7) will indicate the proportions of the various frequencies which might be said to comprise the damped oscillations. Excitation of the circuit at the most dominant of the frequencies in the damped free oscillation would be expected to produce the greatest effect, with a corresponding lower response at the other frequencies. Indeed, the Fourier transform of the damped free oscillation, following the application of an impulse, *is* the frequency response of the system. Maybe this will confirm to the reader the central role of the Fourier theorem in the subject of network analysis.

Summary

We have dealt almost exclusively in previous chapters with the *signal handling* properties of networks, touching only on *power* when discussing the *matching* of circuits one to the other. The power-handling properties of circuits are generaly considered to be more the province of *electrical* rather than *electronic* engineering. However, details are included here to provide some of the essential definitions to be found in the terminology of signal-handling and to give further illustration of the impedance techniques we have learned earlier.

We first confirmed the familiar expression for d.c. power, p_{dc}, developed in a purely resistive circuit: $p_{dc} = vi$ (alternatively $p_{dc} = v^2/R$ or $i^2 R$) and then extended this to include a.c. sources. In the latter case it is only sensible to discuss *average* values over, ideally, an indefinite time-span. The average power for an a.c. source of amplitude v_0 which varies sinusoidally with time is $\frac{1}{2}v_0^2/R$ or $\frac{1}{2}i_0^2 R$. Comparing this average power to that developed by a d.c. source leads to the definition of r.m.s. values of voltage and current, as the voltage or current of a d.c. source which would develop the same power in the load: $v_{rms} = v_0/\sqrt{2}$ and $i_{rms} = i_0/\sqrt{2}$.

Further extending the discussion to include networks with capacitance and inductance we find that the mean power may no longer be obtained from a simple product of voltage and current as these are no longer in phase, a situation to be anticipated from the presence of impedance rather than pure resistance. A calculation of the actual power developed in the network must allow for the phase angle difference, δ, between voltage and current phasors and does so in the so-called *power factor*, $\cos\delta$.

Drawing on our knowledge of complex numbers acquired earlier we defined a current–voltage product, the *complex power*. The real and imaginary parts of the complex power are, respectively, the *real power*, the power actually developed in the resistive part of the impedance, $V_{rms}I_{rms}\cos\delta$, measured in *watts* and the *reactive power*, $V_{rms}I_{rms}\sin\delta$ (not really a power at all), measured in *volt-amps reactive, VAR*. The magnitude of the complex power is the *apparent power*, the current–voltage product with no allowance for phase difference. It is specified simply in *volt-amps*. The apparent, real and imaginary powers can be presented as three sides of a right-angled triangle, the *power triangle* with the apparent power as the hypotenuse. The power triangle is geometrically identical to the *impedance triangle* for the impedance in which the powers are developed.

The ratio of energy stored in an LCR circuit to the rate of energy dissipation at the resonant frequency was found to be the same as a simple ratio of component values. When translated to a ratio of stored energy to energy lost per cycle this was seen to be identical to the Q factor defined in earlier chapters. It is apparent that narrow bandwidth is related to low loss in a tuned circuit and conversely wide bandwidth to high energy loss.

Problems

9.1 In a thyristor full-wave power control circuit, for use at 50 Hz, conduction starts at a point in each half-cycle specified by an angle α (the firing angle) and continues for the remainder of the half-cycle specified by an angle β (the conduction angle) such that $\alpha + \beta = \pi$. Obtain an expression for the r.m.s. voltage in terms of the peak voltage (v_0) and the angle α (that is, find the average value of v_0^2 over a half-cycle and then take the square root). Check that when $\alpha = 0$ the expression obtained is as in equation (9.7).

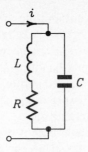

Figure 9.8

9.2 A particular circuit has an impedance:

$$Z = (173.2 + j100)\Omega$$

What is the power factor? Determine the type and value of the component required to effect complete power factor correction for this circuit at 50 Hz.

9.3 Obtain an expression, in terms of the current in the separate branches, for the complex power developed in the network shown in Figure 9.8. If $L = 1$ H, $C = 10\ \mu$F and $R = 10\ \Omega$ find the frequency at which the reactive power is zero.

Appendix A: Coefficients in the Fourier Series

To evaluate the zero order coefficient in the Fourier series we multiply equation (7.1) by dx and integrate from $-\pi$ to π. Since

$$\int_{-\pi}^{\pi} \cos nx \, dx = \int_{-\pi}^{\pi} \sin nx \, dx = 0 \quad \text{for } n = 1, 2, 3, \dots$$

we find that:

$$\int_{-\pi}^{\pi} f(x) \, dx = a_0 \pi$$

Hence
$$a_0 = \frac{1}{\pi} \int_{-\pi}^{\pi} f(x) \, dx$$

A proof that the remaining coefficients are as in equations (7.2) and (7.3) relies on the following identities:

$$\int_{-\pi}^{\pi} \sin mx \cos nx \, dx = 0, \quad \text{for all } m \text{ and } n$$

$$\int_{-\pi}^{\pi} \cos mx \cos nx \, dx = 0, \quad \text{for all } m \neq n$$

$$= \pi, \quad \text{for } m = n$$

$$\int_{-\pi}^{\pi} \sin mx \sin nx \, dx = 0, \quad \text{for all } m \neq n$$

$$= \pi, \quad \text{for } m = n$$

The first of these (identify it as I_1) may be proved by adding and subtracting $\frac{1}{2} \sin nx \cos mx$ within the integral, so that:

$$I_1 = \int_{-\pi}^{\pi} \frac{1}{2} \{ \sin mx \cos nx + \sin nx \cos mx$$

$$+ \sin mx \cos nx - \sin nx \cos mx \} \, dx$$

$$= \int_{-\pi}^{\pi} \frac{1}{2} \{ \sin(m+n)x + \sin(m-n)x \} \, dx$$

$$= \int_{-\pi}^{\pi} \frac{1}{2} \{ \sin px + \sin qx \} \, dx \quad \text{where } p = m+n \text{ and } q = m-n$$

Because, as already stated, the integral of the sine function in the interval $-\pi$ to π is zero for any p or q, we have $I_1 = 0$.

In the second case (identify this as I_2), adding and subtracting $\frac{1}{2} \sin mx \sin nx$ leads to the result that, except when $m = n$, all terms are again zero. The non-zero term is

$$I_2 = \int_{-\pi}^{\pi} \cos^2 nx \, dx$$

$$= \int_{-\pi}^{\pi} \frac{1}{2} (\cos 2nx + 1) \, dx$$

The integral of $\cos 2nx$ in the interval $-\pi$ to π is zero so that $I_2 = \pi$. By similar reasoning the third identity evaluates to π.

To evaluate the nth order coefficients we multiply equation (7.1) by $\cos mx\, dx$ and integrate from $-\pi$ to π. Using the above identities as appropriate we find that:

$$\int_{-\pi}^{\pi} f(x)\cos nx\, dx \;=\; a_n\pi$$

Hence
$$a_n \;=\; \frac{1}{\pi}\int_{-\pi}^{\pi} f(x)\cos nx\, dx$$

so establishing equation (7.2). Multiplying equation (7.1) by $\sin mx\, dx$ and integrating from $-\pi$ to π we may similarly establish equation (7.3), again by the appropriate use of identities proved above.

Appendix B: General Solution of the First-order Linear Differential Equation

We take as an example the case of the RC circuit with a simple harmonic input. A solution is therefore required of:

$$CR\frac{dy}{dt} + y = \cos\omega t$$

Dividing throughout by CR and then multiplying by $e^{\frac{1}{CR}t}$ allows the equation to be integrated, as shown in section 4.4, and the result to be given in the form of equation (4.16):

$$y = ce^{\frac{-1}{CR}t} + e^{\frac{-1}{CR}t}\int e^{\frac{1}{CR}t}\frac{1}{CR}\cos\omega t\, dt$$

Identifying the integral as I we integrate by parts to produce:

$$I = e^{\frac{1}{CR}t}\cos\omega t + \omega\int e^{\frac{1}{CR}t}\sin\omega t\, dt$$

Integrating by parts again:

$$I = e^{\frac{1}{CR}t}\cos\omega t + \omega CR\left\{e^{\frac{1}{CR}t} - \omega\int e^{\frac{1}{CR}t}\cos\omega t\, dt\right\}$$

and the integral I emerges as a term on the right-hand side. Rearranging we have:

$$(1 + \omega^2 C^2 R^2)I = e^{\frac{1}{CR}t}\cos\omega t + \omega CRe^{\frac{1}{CR}t}\sin\omega t$$

so $\quad I = \dfrac{e^{\frac{1}{CR}t}}{1 + \omega^2 C^2 R^2}(\cos\omega t + \omega CR\sin\omega t)$

The complete solution is therefore:

$$y(t) = ce^{\frac{-1}{CR}t} + \frac{1}{1 + \omega^2 C^2 R^2}(\cos\omega t + \omega CR\sin\omega t)$$

or $\quad y(t) = ce^{\frac{-1}{CR}t} + \dfrac{1}{\sqrt{1 + \omega^2 C^2 R^2}}(\cos\delta\cos\omega t + \sin\delta\sin\omega t)$

where $\quad \cos\delta = \dfrac{1}{\sqrt{1 + \omega^2 C^2 R^2}}\quad$ and $\quad \sin\delta = \dfrac{\omega CR}{\sqrt{1 + \omega^2 C^2 R^2}}$

so that $\quad \tan\delta = \omega CR$

Therefore

$$y(t) = ce^{\frac{-1}{CR}t} + \frac{1}{\sqrt{1 + \omega^2 C^2 R^2}}\cos(\omega t - \delta)$$

Further, for $y(t) = 0$ at $t = 0$

$$c = \frac{-1}{\sqrt{1 + \omega^2 C^2 R^2}} \cos -\delta$$

and so finally

$$y(t) = \frac{1}{\sqrt{1 + \omega^2 C^2 R^2}} \left[\cos(\omega t - \delta) - e^{\frac{-1}{CR}t} \cos -\delta \right]$$

The general solution includes both the complementary function and the particular integral which would otherwise be obtained separately by the less general methods outlined in section 4.4.

Appendix C: Laplace Operational Transforms

Apart from the transform for time differentiation, given in Chapter 6, other Laplace operational transforms are:

Combination:

$$\mathcal{L}[f_1(t) + f_2(t)] = F_1(s) + F_2(s)$$

Scaling:

$$\mathcal{L}f(at) = \frac{1}{a}F\left(\frac{s}{a}\right) \quad a > 0$$

Exponential multiplication (frequency shifting):

$$\mathcal{L}[e^{at} f(t)] = F(s - a)$$

Time shifting:

$$\mathcal{L}[f(t - T) \, u(t - T)] = e^{-sT} F(s)$$

Time integration:

$$\mathcal{L}\left[\int_{0-}^{t} f(\tau) \, d\tau\right] = \frac{F(s)}{s}$$

Frequency differentiation:

$$\mathcal{L}[-t f(t)] = \frac{dF(s)}{ds}$$

Frequency integration:

$$\mathcal{L}\left[\frac{f(t)}{t}\right] = \int_{s}^{\infty} F(s) \, ds$$

Two theorems concerning the function and its transform are usually also included in the list of properties. The first follows from the fact that the complete solution of the nth order linear differential equation requires knowledge of the initial value of the derivatives of the function up to order $n - 1$. However, *in the case of zero initial stored energy*, the complete solution is provided by taking the inverse transform of:

$$Y(s) = H(s) \, X(s)$$

It follows that the initial value of the function and all its derivatives must be available in its transform. The ability to find the initial value without evaluating the whole transform is given in the *initial value theorem*:

$$\lim_{t \to 0} f(t) = \lim_{s \to \infty} sF(s)$$

To find the initial value of any particular derivative we write the last equation in the form:

$$\lim_{t \to 0} f(t) = \lim_{s \to \infty} s[\mathcal{L}f(t)]$$

differentiate to the required order and use the operational transform for differentiation to obtain the right-hand side.

As the converse to the initial value theorem, the final value of the output may also be found without evaluating the whole transform and is given by the *final value theorem*:

$$\lim_{t \to \infty} f(t) = \lim_{s \to 0} sF(s)$$

Appendix D: Recommended Reading

[1] Chen, W-K. (1983) *Linear Networks and Systems*, Brooks/Cole Engineering Division. A matrix (state variable) approach to the treatment of linear systems with much reliance on the use of graph theory and directed towards an implementation of the equations in software.

[2] Churchill, R.V. (1958) *Operational Mathematics*, McGraw-Hill. Useful background reading to the development of operational methods pointing to other biographical details on Heaviside (Cooper, 1952) and a detailed definition of the Laplace transform.

[3] Churchill, R.V. (1963) *Fourier Series and Boundary Value Problems*, McGraw-Hill. A clear definition of the Fourier series and its application.

[4] Close, C.M. (1966) *The Analysis of Linear Circuits*, Harcourt Brace and World. A comprehensive text which includes much detail on the classical methods for the analysis of circuits together with an extensive treatment of the use of the complex variable.

[5] Cooper, J.L.B. (1952) *Math. Gazette*, vol. 36, pp. 5-19. A detailed account of the controversy surrounding the work of Oliver Heaviside.

[6] Lanczos, C. (1966) *Discourse on the Fourier Series*, Oliver and Boyd. Interesting background details to the development of the Fourier series.

[7] Nilsson, J.W. (1986) *Electric Circuits*, Addison-Wesley. A large book covering much the same area as the present volume but with heavy reliance on problem working.

Appendix E: Answers and Guidance for Problems

Chapter 1

1.1 (a) Non-linear (b) Linear (c) Non-linear

1.2

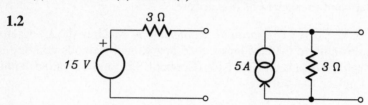

1.3

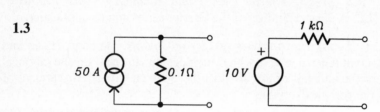

1.4 At 0.01 Hz the amplitude remains unity, at 1 Hz the amplitude is reduced to about 1/10.

Chapter 2

2.2 (a) $v_1 = 1.6$ V (b) $v_1 = -0.4$ V

2.4 Left node 10 V, right node 8 V; mesh currents in clockwise sense from left to right, 3 A, 0.5 A and -0.5 A

2.5 $v_N = 2.8$ V, $i_1 = 0.6$ A and $i_2 = 0.4$ A

Chapter 3

3.1 $v_0 = 8$ V

3.2 $v_0 = 2.8$ V

3.3

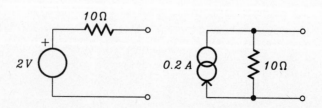

3.4 A source transformation applied to the 3 A source and shunt resistor is a helpful start.

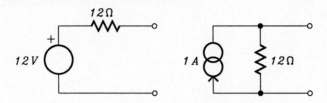

Chapter 4

4.1
$$R\frac{dq}{dt} + \frac{1}{C}q = x(t)$$

Before $t = 0$ there is no current flowing and the output voltage is zero. At the instant the input is replaced by a short circuit the output drops to $-v_0$ from which it recovers to zero as $y(t) = -v_0 e^{-t/RC}$.

4.2
$$\frac{dy}{dt} + \frac{1}{CR}y = \frac{dx}{dt}$$

Follow the procedure in Example 4.4 to obtain $y_p(t) = a\cos(\omega t - \delta)$ with
$$a = \frac{\omega CR}{\sqrt{1 + \omega^2 C^2 R^2}}$$
and $\tan\delta = \omega CR$. Note: the input is $\sin\omega t$ so there is a total phase shift from input to output of $(\pi/2 - \delta)$. The complementary function (solution of the homogeneous equation) is as given in Example 4.4.

4.3
$$\frac{d^2y}{dt^2} + \frac{R}{L}\frac{dy}{dt} + \frac{1}{LC}y = \frac{R}{L}\frac{dx}{dt}$$

4.4 The answer may be obtained by considering the triangles with vertex angle δ and $N\delta$ and it is helpful to introduce the parameter r. For zero amplitude $N\delta = 2n\pi$, n an integer.

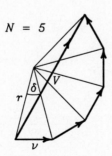

$N = 5$

4.5

$$Ri_1 + L_1\frac{di_1}{dt} + M\frac{di_2}{dt} = v(t)$$

$$Ri_2 + L_2\frac{di_2}{dt} + M\frac{di_1}{dt} = 0$$

Chapter 5

5.1 This is the same problem as 4.4 but solved using complex exponentials.

$$
\begin{aligned}
V &= \left| v \sum_{n=0}^{N-1} e^{jn\delta} \right| \\
&= \left| v \frac{1 - e^{jN\delta}}{1 - e^{j\delta}} \right| \\
&= \left| v \frac{e^{j\frac{1}{2}N\delta}}{e^{j\frac{1}{2}\delta}} \frac{e^{-j\frac{1}{2}N\delta} - e^{j\frac{1}{2}N\delta}}{e^{-j\frac{1}{2}\delta} - e^{j\frac{1}{2}\delta}} \right|
\end{aligned}
$$

Hence the answer.

5.2 The free response is given by $y = 2e^{-\sigma t} \cos \omega_m t$ where $\sigma = \dfrac{R}{2L}$ and

$$
\omega_m = \sqrt{\frac{1}{LC} - \frac{R^2}{4L^2}} = \omega_0 \sqrt{1 - \frac{\omega_0^2 C^2 R^2}{4}} \quad \text{with } \omega_0^2 = \frac{1}{LC}.
$$

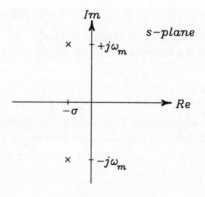

5.3 $\dfrac{L}{R}\dfrac{dy}{dt} + y = v(t)$ $LC\dfrac{d^2y}{dt^2} + CR\dfrac{dy}{dt} + y = v(t)$

$$
H(s) = \frac{R/L}{s + R/L} \qquad H(s) = \frac{1/LC}{s^2 + sR/L + 1/LC}
$$

5.4 Follow the procedure in Examples 5.4 and 5.5 to obtain the same results as for 5.3 above.

5.5 $H(s) = \dfrac{sL}{2sL + R}$

Chapter 6

6.2 The solution is very straightforward and not even partial fraction expansion is required: $y(t) = u(t)e^{-\alpha t}$ where $\alpha = 1/CR$.

6.3 $y(t) = Ku(t)\left[1 - e^{-\alpha t}\right]$ where $K = \dfrac{R_2}{R_1 + R_2}$ and $\alpha = \dfrac{1}{CR_p}$ with $R_p = \dfrac{R_1 R_2}{R_1 + R_2}$

6.4 Perform only a 'partial' partial fraction expansion with one first-order and one second-order denominator. The inverse transform then produces the result exactly as given in Example 4.4.

Chapter 7

7.1 The spectrum of the pure sine wave will be a single sharp line as in (a). When the sine wave is lightly damped the function is no longer pure, that is, it is no longer a single oscillation, and the spectrum can no longer be sharp. We might expect this to be represented by a broadening of the formerly sharp line spectrum as in (b).

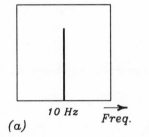

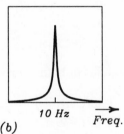

(a) *10 Hz* → *Freq.*   (b) *10 Hz* → *Freq.*

7.2 (a) Component at $2\omega_0$, a cosine term, has an amplitude of $-4/3\pi$ (ω_0 is the fundamental frequency).

(b) Subtracting $\sin\omega_0 t$ from the half-wave rectified signal produces the waveform in (a). So the lowest harmonic is at frequency ω_0 (the fundamental frequency) and amplitude 1.

7.3 The function is odd, so only sine terms will be required in its Fourier series expansion. The d.c. level ($a_0/2$) is 0.5 V, so $a_0 = 1$. A component at the fundamental frequency, ω_0, is present with amplitude $1/\pi$.

7.4 $F(j\omega) = \dfrac{T}{2}\dfrac{\sin^2 \omega T/4}{(\omega T/4)^2}$

7.5 Follow exactly the same procedure as adopted for the case of the delayed unit impulse.

Chapter 8

8.1 $40 - 12 + 20 = 48$ db

The gain = antilog $(48/20) = 251$, whereas by direct calculation one would find 250.

8.2 $$Z = \frac{R}{1 + j\omega CR} + j\omega L$$

Using the given values in the expression after rationalization produces: $Z = 500 + 500j$. $|Z| = 500\sqrt{2}$ and as $v_0 = 1$, $i_0 = 1/(500\sqrt{2}) = 1.414$ mA, $\tan \delta = 1$ and so $\delta = 45°$.

8.3 The load impedance reflected into the primary is $3/n^2$ (n is the turns ratio, secondary to primary). For matching: $3/n^2 = 1875$ so the turns ratio, primary to secondary $(1/n)$ is 25.

8.4 Assuming zero initial phase for the source voltage, the phasor for the voltage between A and B is:

$$
\begin{aligned}
v_{AB} &= \frac{1}{2} - \frac{1/j\omega C}{R_1 + 1/j\omega C} v_0 \\
&= \frac{1}{2} - \frac{1}{1 + j\omega CR_1} v_0 \\
&= \frac{1}{2} \frac{j\omega CR_1 - 1}{j\omega CR_1 + 1} v_0
\end{aligned}
$$

Taking the modulus of both numerator and denominator we find:

$$|v_{AB}| = \frac{1}{2} \frac{\omega^2 C^2 R_1^2 + 1}{\omega^2 C^2 R_1^2 + 1} v_0$$

which has the value $\frac{1}{2}$ independent of frequency, illustrating the fact that the circuit is an *all-pass filter*. It instead we rationalize the expression:

$$v_{AB} = \frac{1}{2} \frac{(1 - \omega^2 C^2 R_1^2) - 2j\omega CR_1}{1 - \omega^2 C^2 R_1^2} v_0$$

A phase change of $\pi/2$ occurs when the right-hand side is wholly imaginary, that is, when $\omega^2 C^2 R_1^2 = 1$ or $\omega = 1/CR_1$.

8.5 $$H(j\omega) = \frac{R_2}{R_1 + R_2} \frac{1}{1 + j\omega CR_p} \quad \text{where } R_p = \frac{R_1 R_2}{R_1 + R_2}$$

8.6 Using $\omega_0 = 2\pi f_0 = \sqrt{1/LC}$ and $Q = 1/\omega_0 CR$ we find $f_0 = 15.9$ kHz and $Q = 10$.

8.7 $$\omega_m = \omega_0 \sqrt{1 - \frac{1}{2Q^2}}$$

Chapter 9

9.1 $\qquad V_{rms}^2 = \dfrac{1}{\pi} \displaystyle\int_{\alpha}^{\pi} v_0^2 \sin^2\theta \, d\theta$

whence $V_{rms} = v_0 \left[\dfrac{1}{2} - \dfrac{\alpha}{2\pi} + \dfrac{\sin 2\alpha}{4\pi} \right]^{\frac{1}{2}}$

which for $\alpha = 0$ gives the result obtained in equation (9.7).

9.2 $|Z| = 200\ \Omega$ and $\cos\delta = 0.866$ so $\delta = 30°$ (lagging). The circuit admittance is:

$$Y = \dfrac{173.2}{200^2} - j\dfrac{100}{200^2}$$

For power factor correction a shunt component with an admittance of $j100/200^2 = j/400$, that is an impedance of $-j400$, is required. This is a capacitative reactance. Using $1/\omega C = 400$ the value of the capacitor is found to be $7.96\ \mu F$.

9.3 The complex power:

$$\mathcal{P} + jQ = I_1^2 R + j\left\{ I_1^2 \omega L + I_2^2(-1/\omega C) \right\}$$

where I_1 and I_2 are the r.m.s. currents in the branches with the inductor and the capacitor respectively. For zero reactive power:

$$I_1^2 \omega L = I_2^2 \dfrac{1}{\omega C} \quad \text{that is} \quad \dfrac{I_2^2}{I_1^2} = \omega^2 LC$$

But $\dfrac{I_2^2}{I_1^2} = \dfrac{|Z_1|^2}{|Z_2|^2} = \dfrac{R^2 + (\omega L)^2}{(1/\omega C)^2}$

whence $\omega^2 = \omega_0^2(1 - \omega_0^2 C^2 R^2)$ with $\omega_0^2 = 1/LC$

The values given determine that $\omega_0^2 C^2 R^2 = 10^{-3}$ and therefore $\omega = \omega_0$ to a good approximation. A circuit which is power factor corrected (zero reactive power) is a resonant circuit at the chosen frequency which in this case is very close to 50 Hz (50.3 Hz).

Index

active circuit, 3, 5
active element, 45
admittance, 46, 130
algebraic sum, 15
amplitude, 65
    spectrum, 124
analysis, 1
apparent power, 162
Argand diagram, 64–69, 127, 129, 161
argument, 64, 68
auxiliary equation, 69
average power, 156

battery, 3, 5
bel, 138
Bernoulli, 103
binary sequence, 36
Bode, 119
Bode plot, 138, 141
branch, 17

capacitance, 47
capacitor, 3, 41, 45, 47
    electrolytic, 49
    $i - v$ relationship, 4
Cauchy, 80
causal, 110
    exponential, 110
    function, 111
    system, 98, 110
causality, 11
CD, 138
characteristic equation, 69, 72
characteristic function, 102
charge conservation, 15
circuit, 1
    active, 3
    construction, 1
    equivalent, 16, 31, 32
    linear, 2
    non-linear, 2
    passive, 3
CIVIL, 159
coefficient of coupling, 59

coil-former, 58
coils, 58
comb function, 116
combination, linear, 3
common mode interference, 51
compact disc, 138
comparison of coefficients, 92
complementary function, 54, 69, 86, 170
completing the square, 93
complex algebra, 67
complex amplitude, 67
complex conjugate, 65
complex exponential, 66–77, 106, 161
    input, 70, 73, 85, 87
complex frequency, 70, 81
complex impedance, 75, 87
complex number, 63
    cartesian form, 65
    Euler form, 65
    imaginary part, 65, 66
    polar form, 65
    real part, 66
    trigonometric form, 65
complex power, 161
conductance, 130
conservation
    of charge, 15
    of energy, 15
controlled source, 5, 23, 31, 45
convolution, 12
    in frequency domain, 114
    integral, 12, 95
copper losses, 137
corner frequency, 142
cosine function, 51
cotree, 17
coupled equations, 46
coupling, 77
    coefficient, 59
    perfect, 59
cover-up rule, 91
current
    divider, 38, 41
    phasor, 157

short-circuit, 8
current source
    ideal independent, 6
    $i - v$ characteristics, 6
current–voltage relationships, 4
cutsets, 17

d'Alembert, 103
damped oscillations, 164
db, 138
dead source, 9, 31
decade, 142
decibel, 138
demodulation, 51
dielectric, 5
    leakage, 5
differential equation, 5, 63, 73
    first-order, 48, 81, 169
    homogeneous, 55, 68
    linear, 47, 48, 50, 55, 68, 169
    non-homogeneous, 48, 50, 54, 81
    second-order, 50
differential mode, 51
differentiator, 2
digital signal processing, 119
digital signals, 36
digital techniques, 51
Dirichlet, 103
dot convention, 58
driving-point
    admittance, 129
    function, 46, 129
    impedance, 129
dual network, 41

eigen function, 102
electronic engineer, 1
energy
    conservation, 15
    dissipation, 3
    storage elements, 47, 156, 162
engineer, 1
    electronic, 1
equivalent circuit, 16, 31, 32
Euler, 65, 103
    expression, 106
    form, 65, 107, 127
even function, 105
excitation, 1, 45
exponential
    decay, 49
    function, 71
external mesh, 41

f-circuit, 17
farad, 4
Faraday, 4
field theory, 5
filter
    Butterworth, 151
    first-order
        frequency response, 140
        low-pass, 140
    ideal
        band-pass, 120
        low-pass, 118
    line, 51
    second-order
        band-pass, 143
        frequency response, 145
        high-pass, 150
        low-pass, 150
        notch, 148
final value theorem, 172
flux, self-linkage, 57
forcing function, 48, 55, 73
four-terminal network, 45
Fourier, 51, 103
    coefficients, 103–108, 167–168
    integral, 104
    waveform analyzer, 51
Fourier series, 103–109, 167
    complex exponential form, 106–107
    trigonometric form, 103–106
Fourier transform, 99, 109–120, 125, 164
    duality, 113, 115
    inverse, 109–110
    of convolution, 113
    of periodic functions, 116
    of rectangular pulse, 117
    of time-shifts, 114
    of unit impulse, 112
Fourier's theorem, 52, 55, 102–103, 124
free response, 48, 55, 68, 72, 98
frequency, 69
    component, 106
    components, 102
    domain, 102, 106, 107
    spectrum, 112, 125
    synthesis, 51
frequency response, 102, 125–151
    Butterworth, 151
    Euler form, 140
    first-order low-pass, 140
    second-order band-pass, 145
    second-order high-pass, 150
    second-order low-pass, 150

second-order notch, 148
standard form, 146
function
even, 105
odd, 104
periodic, 104
system, 1
fundamental frequency, 106

generalized impedance, *see* Laplacian impedance
graph, 17
theory, 17, 41

harmonic
functions, 51
series, 104
Heaviside, Oliver, 80
Henry, 4
henry, 4

ideal dependent source, 5
ideal independent current source, 6
$i - v$ characteristics, 6
ideal independent source, 5
ideal independent voltage source, 6
$i - v$ characteristics, 6
ideal source, 5
ideal transformer, 137
imaginary axis, 64
imaginary part, *see* complex number, imaginary part
impedance, 46, 75, 86, 124
cartesian form, 129
Euler form, 129
potential divider, 76, 88, 126, 140
reflected, 136
steady a.c., 126–129
triangle, 129, 157
improper fraction, 89
impulse response, 11
L C circuit, 96
RC circuit, 95
independent source, 45
inductance, 47, 57
inductor, 3, 41, 45, 47
$i - v$ relationship, 4
information, 39
initial phase, 128, 157
initial stored energy, 48, 50
initial value theorem, 171
input, 1, 2, 45
admittance, 129
impedance, 129

instantaneous power, 155, 160
integration in the complex plane, 83
integro-differential equations, 26
internal resistance, 7
inverse Fourier transform, 109–110
inverse Laplace transform, 82, 86

joule, 3
joule heating, 3, 15

KCL, *see* Kirchhoff's current law
kernel, 81
Kirchhoff's current law, 15–26, 40, 46, 132
Kirchhoff's laws, 86
in *s*-plane, 82
Kirchhoff's voltage law, 15–26, 40, 46, 47, 50, 59, 75, 130
KVL, *see* Kirchhoff's voltage law

L C circuit, 85
L CR circuit, 93
Laplace, 80
Laplace transform, 81, 109, 124
bilateral, 99, 110
final value theorem, 172
frequency differentiation, 171
frequency shifting, 171
initial value theorem, 171
inverse, 82, 86
inversion integral, 82
of frequency integration, 171
of functions, 82
of integration, 82, 171
of sum, 82, 171
of time-shifts, 82, 171
operational, 171
pairs, 82
product, 95
scaling, 82, 171
unilateral, 99
Laplacian impedance, 87
LC circuit, 50, 69
LCR circuit, 26, 76, 162
Lenz, 4
Lenz's law, 59
line filter, 51
linear circuit, 2
linear combination, 3, 73
linear differential equation, 47, 48, 50, 55, 169
logarithmic scale, 137
long division, 89
loop equations, 20

loop-current method, *see* loop equations
lumped parameter, 5

magnetic tape, 138
mains filter, 51
mains voltage, 157
mass, 47
matrix algebra, 18
maximum power theorem, 39
mean power, 156, 160, 162
mechanical system, 47
mesh, 41
mesh equations, 20, 40, 46
mesh-current method, *see* mesh
        equations
method of undetermined coefficients, 55
microphones, 5
modulation, 120
modulus, 64, 68
motor-car suspension, 47
music, 51
mutual inductance, 58

n-terminal network, 45
network, 1
    duals, 40
    electrical, 66
    function, 71
node, 17, 41
node equations, 20, 40, 46
node-voltage method, *see* node equations
noise power, 39, 138
non-ideal independent source, 7
non-zero initial stored energy, 97
Norton equivalent
    circuit, 37
    conductance, 38
    resistance, 38
Norton's theorem, 33, 41
Nyquist, 119

octave, 142
odd function, 104
ohm, 4
Ohm's law, 4, 155
one-port, 46, 129
open circuit, 7, 31, 41
open-circuit voltage, 7
operational methods, 80, 124
output, 1–2, 45

parallel
    connection, 41
    LC, 133

RC, 132
    resonance, 144
partial fraction expansion, 89–92
    comparison of coefficients, 92
    completing the square, 93
    cover-up rule, 91
particular integral, 71, 86, 170
particular solution, 54
passive circuit, 3
passive sign convention, 4, 6
peak voltage, 157
peak-to-peak voltage, 157
perfect coupling, 59
period, 105
periodic extension, 104
periodic function, 104
phase, 53
    angle, 53, 65
    difference, 53, 129
    shift, 52
    spectrum, 124
phasor, 66, 67, 71, 107, 127, 161
    diagram, 66
photocells, 5
physical system, 1
planar, 20
polar form, 65
pole, 72
    diagram, 72
pole-zero diagram, 73
polynomial, 72
potential divider, 20, 31–41, 76, 88
power, 155–162
    amplifier, 5
    apparent, 162
    average, 156
    complex, 161
    instantaneous, 155
    mean, 156, 160, 162
    ratio, 138
    reactive, 160, 162
    real, 162
    supply, 5, 39, 45
    transfer, 39
    transmission, 51
power factor, 158
    correction, 159
power triangle, 161
practical source, 7
pre-amplifier, 5
primary winding, 60
principle of superposition, 2, 30–35, 102
    in *s*-plane, 82
proper fraction, 89

Q, 147, 164
quadrature, 53
quality factor, 76, 147

R–2R ladder, 36
r.m.s., *see* root-mean-square value
radio antennae, 5
radio frequencies, 51
RC circuit, 68, 84, 90, 169
reactance, 129
    capacitative, 129
    inductive, 129
reactive power, 160, 162
real axis, 64
real part, *see* complex number, real part
real power, 162
reference direction, 4, 6
reflected impedance, 136
residue theorem, 83
resistance, internal, 7
resistor, 3
    four-terminal, 46
    $i - v$ relationship, 4
resistors
    parallel, 16
    series, 16
response, 1, 45
root-mean-square value, 157
rotating vector, 52

s-plane, 69, 72, 81
s.h.m., *see* simple harmonic motion
sampling, 116
sampling theorem, 119
second-order system, 50
secondary winding, 60, 77
self-inductance, 58
series
    LC, 133
    RC, 130
    resonance, 144
series connection, 41
Shannon, 119
short circuit, 7, 31, 41
short-circuit current, 8
side-bands, 120
sign convention, passive , 4
signal
    generator, 51
    power, 39, 138
    source, 51
signal-to-noise ratio, 138
simple harmonic motion, 50, 65
sinc function, 117

sine function, 51, 102
singularity functions, 48
source, 5
    controlled, 5
    current, *see* current source
    ideal, 5
    ideal dependent, 5
    ideal independent, 5
    non-ideal independent, 7
    practical, *see* practical source
    resistance, 7
    transformation, 9, 37, 97
    voltage, *see* voltage source
spectra, 106
square wave, 52
stability, 11
state equations, 27
state of system, 27
state vector, 27
steady a.c., 124–126, 155
    impedance, 126–129
sub-system, 2
superconductivity, 50
superposition, 37
superposition principle, *see* principle of
    superposition
suppressed source, 31
susceptance, 130
switching transient, 55
synthesis, 1, 124
system, 1
    differential equation, 47, 48, 73, 86,
        124
    function, 1, 71, 85, 88
    physical, 1, 66
    time-invariant, 102

television, 51
Tellegen's theorem, 25
terminal characteristic, 7
Thevenin
    equivalent, 34, 36, 39
        resistance, 36
Thevenin's theorem, 33, 41, 88
time domain, 81, 102, 106, 107
transducer, 5, 39
transfer function, 45, 129
transformer, 58, 77, 135
    dot convention, 58
    ideal, 137
    step-down, 60
    step-up, 60
    unity-coupled , 77
transistor, 3, 5

tree, 17
trial function, 69
turns ratio, 60, 77
two-port, 129
    network, 45
    parameters, 46

undetermined coefficients, 71
unit
    impulse, 11, 94
      response, 94
    step function, 48, 90–94
unity-coupled transformer, 60, 136

VA, *see* volt-amps
VAR, *see* volt-amps reactive
vector algebra, 53
volt-amps, 162
volt-amps reactive, 162

voltage
    base-emitter, 5
    drop, 4
    open-circuit, 7
    phasor, 157
    rise, 4
voltage source
    ideal independent, 6
      $i - v$ characteristics, 6

watts, 162
waveform analyzer, 51
waveform generator, 51
Wheatstone bridge, 16, 18, 37
work, 155

zero initial stored energy, 85, 90
zero input response, 98
zero state response, 86, 98
ZIR, *see* zero input response
ZSR, *see* zero state response

STOCHASTIC DIFFERENTIAL EQUATIONS & APPLICATIONS

XUERONG MAO, Reader in Mathematics, Department of Statistics and Modelling Science, University of Strathclyde

ISBN 1-898563-26-8 (1997) 384 pages

This introduction to stochastic differential equations covers basic principles, with much on theory and applications not previously available in book form. It is an advanced course text (and also a reference source) for pure and applied mathematicians, statisticians, engineers in control and communications, and other scientific areas of application and research.

The text covers generalised Gronwall inequality and Bihari inequality ... introduces Brownian motions and stochastic integrals ... analyses classical Ito formula and the Feynman-Kac formula ... demonstrates manifestations of the Lyapunov method and Ruzumikhin technique ... discusses Cauchy-Marayama's and Carathedory's approximate solutions to stochastic differential equations ... and more.

CALCULUS Introduction to Theory and Applications in Physical and Life Science

R.M. JOHNSON, Department of Mathematics and Statistics, University of Paisley

ISBN: 1-898563-06-3 (1995) 336 pages £15.00

This7 lucid and balanced text for first year undergraduates (UK) conveys the clear understanding of the fundamentals and applications of calculus, as a prelude to studying more advanced functions. Short and fundamental diagnostic exercises at chapter ends testing comprehension, before moving to new material.

LINEAR DIFFERENTIAL AND DIFFERENCE EQUATIONS:
A Systems Approach for Mathematicians and Engineers

R.M. JOHNSON, Department of Mathematics and Statistics, University of Paisley

ISBN: 1-898563-12-8 (1996) 200 pages

This text for advanced undergraduates and graduates reading applied mathematics, electrical, mechanical, or control engineering employs block diagram notation to highlight comparable features of linear differential and difference equations, a unique feature found in no other book. The treatment of transform theory (Laplace transforms and z-transforms) encourages readers to think in terms of transfer functions, i.e., algebra rather than calculus. This contrives short-cuts whereby steady-state and transient solutions are determined from simple operations on the transfer functions.

SIGNAL PROCESSING IN ELECTRONIC COMMUNICATIONS
MICHAEL J. CHAPMAN, DAVID P. GOODALL, and NIGEL C. STEELE,
School of Mathematics and Information Sciences, Coventry University.

ISBN 1-898563-30-6 (1997) 288 pages

This text for advanced undergraduates reading electrical engineering, applied mathematics, and branches of computer science involved with signal processing (speech synthesis, computer vision and robotics).will also serve as a reference source for graduate workers in academia and industry.

Signal processing is an important aspect of electronic communications in its role of transmitting information, and the mathematical language of its expression is developed here in an interesting and informative way, imparting confidence to the reader.

The first part of the book focuses on continuous-time models, and contains chapters on signals and linear systems, and on system responses. Fourier methods, so vital in the study of information theory, are developed prior to a discussion of methods for the design of analogue filters. The second part of the book is directed towards discrete-time signals and systems. There is full development of the z- and discrete Fourier transforms to support the chapter on digital filter design.

All preceding material in the book is drawn together in the final chapter on some important aspects of speech processing which provide an up-to-date example of the use of the theory. Topics considered include a speech production model, linear predictive filters, lattice filters and cepstral analysis, with application to recognition of non-nasal voiced speech and formant estimation.

Prerequisites are simply an elementary knowledge of algebra (e.g. partial fractions), and also calculus including differential equations. A knowledge of complex numbers and of the basic concept of a function of a complex variable is also needed.

Contents: Signal and Linear System Fundamentals; System Responses; Fourier Methods; Analogue Filters; Discrete-time Signals and Systems; Discrete-time System Responses; Discrete-time Fourier Analysis; The Design of Digital Filters; Aspects of Speech Processing; Appendices: The Complex Exponential; Linear Predictive Coding Algorithms; Answers.

FINITE ELEMENT TECHNIQUES IN STRUCTURAL MECHANICS

CARL T.F. ROSS, Department of Mechanical and Manufacturing Engineering, University of Portsmouth

ISBN: 1-898563-25-X (1997) 224 pages

Written in a step-by-step methodological approach, this undergraduate and post-graduate text will serve for courses in mechanical, civil, structural and aeronautical engineering, and naval architecture.

"All Carl Ross' previous books are very clear and well written. This thoroughly interesting text is no exception. The worked examples and practice problems are particularly useful. I will continue to recommend Professor Ross' books to my students. For anyone requiring an introduction to finite element analysis, this text is excellent." - *Journal of Strain Analysis* (Dr. S.J. Hardy, University College, Swansea)

FINITE ELEMENT PROGRAMS IN STRUCTURAL ENGINEERING AND CONTINUUM MECHANICS

CARL T.F. ROSS, Department of Mechanical and Manufacturing Engineering, University of Portsmouth

ISBN 1-898563-28-4 (1996) 650 pages

This undergraduate and postgraduate text will serve for courses in mechanical, civil, structural and aeronautical engineering; and naval architecture. A step-by-step methodological approach.

"Computer programs for finite element analysis ... students and lecturers may find them of value" - *The Structural Engineer* (Professor I.A.Macleod, Strathclyde University, Glasgow)

"These programs, written in Quick Basic, utilize the finite element method to solve a variety of engineering problems ranging from static and dynamic analysis in two and three dimensions to two-dimensional field ... recommended to readers with a strong theoretical background, and upper-division undergraduates through to professionals" - *Choice*, American Library Association (D.A. Pape, Alfred University, USA)

DYNAMICS OF MECHANICAL SYSTEMS

CARL T.F. Ross, Department of Mechanical Engineering and Manufacture, University of Portsmouth

ISBN: 1-898563-34-9 (1997) 260 pages

This fundamental introduction for first and second year undergraduates reading mechanical, civil, structural and aeronautical engineering, and naval architecture will also appeal BTEC students. The methodical approach is aimed at students who find difficulty with mathematics and Newtonian Physics.